The Smell Book

The Smell Book
Scents, Sex, and Society

Ruth Winter

J. B. LIPPINCOTT COMPANY
Philadelphia and New York

Copyright © 1976 by Ruth Winter
All rights reserved
First Edition

U.S. Library of Congress Cataloging in Publication Data

Winter, Ruth, birth date
  The smell book.

  Bibliography: p.
  Includes index.
  1. Smell. 2. Pheromones. 3. Perfumes. I. Title.
  QP458.W56   152.1'66   76-18733
  ISBN-0-397-01163-6

To ARTHUR—
who makes my whole world smell sweet

Contents

Acknowledgments   9

PART I: SCENT LANGUAGE

1 Smell Signatures   13
2 The Supersense   20
3 The Scent of Sex   33
4 Sickly Smells   57
5 When the Nose Doesn't Know   73

PART II: THE SUCCESS OF SWEET SMELLS

6 Perfume Politics   87
7 The Scent Manipulators   105
8 Malodor Maladies   125
9 The Human Use of Common Scents   141
   Bibliography   159
   Index   171

Acknowledgments

THE AUTHOR wishes to thank the many experts who gave of their time and knowledge, in particular: Lee Horn and Darrel Huebner of the 3M Company; Lynn Waplington and Jane Barr of Burson-Marsteller; Peter Midwood, Hugh Watkins and Juanita Byrne-Quinn of Proprietary Perfumes, Ltd.; Annette Green of the Fragrance Foundation; Ernest Shiftan of International Flavors and Fragrances; Richard P. Michael, Ph.D., of Emory University; Dr. Richard Doty and Dr. George Preti of the Monell Chemical Senses Center; Dr. Robert I. Henkin of Georgetown University; Dr. Andrew Dravnieks of the Illinois Institute of Technology; Dr. J. E. Amoore of the U.S. Department of Agriculture's Western Regional Research Laboratory; and Dr. A. A. Schleppnik of Monsanto Flavor/Essence, Inc.

Part I:
Scent Language

1 Smell Signatures

NO MATTER HOW we scrub and clean ourselves, we all emit a unique individual odor. Furthermore, we are all profoundly affected by other people's odors and by the odors in our environments. No aspect of our behavior is immune. We communicate with a silent, invisible, often subliminal smell language in our bedrooms, dining rooms, offices—wherever we are.

Our ability to receive smell messages accompanies us into the world at birth. Before our other senses are fully operative, we are receiving survival information through our noses, which tell us where and who our mothers are and the location of the food they provide.

The sense of smell gives us sexual, gustatory, and psychological pleasure. It stimulates our memories and remains faithful to us long after the other senses have dimmed. Our olfactory sense functions when the other senses do not. We need light for sight and the direct application of molecules to the tongue for taste. When we sleep, our senses of hearing and touch are partly turned off but our noses are ever vigilant. And yet, we are not proud of our sense of smell. We brag about our twenty-twenty vision and our fine palates, although we can taste only four things and flavor is largely aroma. We tout our keen sense of hearing, but we do not boast about our ability to smell.

We don't even have an adequate vocabulary for our sense of smell. The word "smell" is confusing, as it serves as both a verb and a noun. And we have no names for specific odors; we say only that they "smell like" something or other.

Why are we so self-conscious about our ability to smell?

First of all, it reminds us that we are animals. It's true that we don't go around sniffing each other quite as obviously as dogs and rats, but you have only to watch a human mother sniffing the head of her infant to realize how instinctive smell behavior is; and you have only to consider how you react to the scent of someone you love to recognize how smells affect us socially.

In a fascinating experiment illustrating social smell behavior, researchers from the University of California tested scent and the use of personal space, using male and female "stimulus persons" at an amusement park. When the experiment participants wore perfume or after-shave lotion, the individuals standing in line close by them moved farther away than when no scents were used. The "stimulus persons" apparently were repellent to others in their environment despite the fact that the perfume and after-shave lotion they wore were popular, pleasant scents. Evidently, the desire to protect one's personal space from scent stimulation emanating from strangers is unconscious but irresistible. This behavior is amazingly similar to that of animals in the wild when a strange member of the species is introduced into their home territory.

In our culture a correlation between scent and personal space is reflected in our language. What do we call a disagreeable person when we want to warn someone to stay away from him or her? A "stinker" or a "skunk."

We know that in our society one can't have bad breath, sweaty underarms, or noticeable genital odor. You can tell people they need a haircut or to wash their face, but if you tell them they smell, you are really insulting. The height of crudeness is the passing of gas in public. Sociologists call it the "fart taboo"—which brings us to the second major reason we are embarrassed about our ability to smell.

Sigmund Freud pointed out that in earliest infancy there is no trace of shame about the excretionary functions or disgust at their products. Small children show great interest in the excreta of their own bodies. Indeed, until the advent of "civilization," adults did too. Feces were used to renew the earth, as they still are in many countries, notably China, and urine was employed as a detergent for washing. (We still use one of the major components of urine, ammonia, in our cleaning products, but it costs a lot more.)

As discoveries about the relationship between dirt and disease developed, and soap came into general use, we began to dislike anything associated with waste products. We imposed restrictions on our children. They not only had to control the time and place of their eliminations; they had to keep them secret. They were forced to learn to be ashamed and disgusted about their own and other people's natural functions. Thus, the strong, telltale odor of fecal wastes became associated with a sense of shame.

Freud maintained that this reaction to excrement goes far beyond the bounds of rationality, and he even considered the consequent repression of the sense of smell as a major cause of mental illness.

"With the assumption of an erect posture by man

and with the depreciation of his sense of smell," Freud wrote in *Civilization and Its Discontents* published in 1930, "it was not only his anal eroticism which threatened to fall a victim of organic repression but the whole of his sexuality."

Freud and many of his disciples have pointed out that, because the genitalia are closely associated with excretion and there is a characteristic smell surrounding the organs, embarrassment about such smells and shame and inhibition about sex are intertwined.

In fact, according to Richard von Krafft-Ebing, the nineteenth-century German neurologist, the rise in the use of the bathtub coincided with the increased incidence of smell-related fetishes. He said the popularity of the handkerchief, shoe, and underclothing, as well as feet, sweat, and excrement in sexual fetishes was partly due to the pungent bodily odors associated with them. Such aberrations were counterreactions to the cultural suppression of the sense of smell.

Some social scientists believe another contributing factor to the repression of our olfactory sense was the change from breast to bottle feeding. They maintain that the frustration of the instinctive search by infants for the soft, aromatic breast that has been replaced by the sterile, odorless bottle inhibits the normal development of pleasurable reactions to healthy body odors.

The evidence seems clear that the neurotic repression of our instinctive behavioral reactions to smell is the cause of the dramatic odor threshold changes seen at various ages. It has been shown repeatedly that there is a sharp drop in a child's ability to detect odors around the age of ten, the beginning of social awareness and conformity, and then again at eighteen, the age of entrance into adulthood.

This self-consciousness about our own and others' body odors is fed constantly today by television, newspaper, and magazine advertisements. We are literally told that we stink—our mouths, our armpits, and our genitals need special products to make them and us socially acceptable. As a result of this obsession with cleanliness and odorlessness, we have done our best to repress smells in our world. Billions and billions of dollars' worth of vented bathrooms, household and body deodorants, perfumes and other antismell devices have been developed. And yet the most basic of our senses remains magnificently intact, ready to inform us. We have tried to push its messages into oblivion, but they are still there, as powerfully stimulating as ever.

Our own odors and the odors surrounding us cannot be hidden. Each of us has a unique olfactory system. We have slightly different combinations of odor receptors and diverse individual responses associated with various smells, and our own odor depends on body metabolisms that are not identical. Our ability to smell and how we ourselves smell vary with the time of day and, for women, with the time of month as well. Therefore, how we ourselves smell and how we smell someone else is strictly personal. If you doubt this, just consider that bloodhounds can identify each of us by our scent in a crowd and can follow our individual odor trails for as long as two weeks after we have invisibly imprinted them.

Bloodhounds may surpass us in tracking scent, but our own sense of smell is amazingly sensitive. We can detect one part of an odorant of natural gas in fifty million parts of air. Our olfactory ability is ten thousand times more sensitive than our sense of taste. Furthermore, olfaction, among all our various senses, is the one with the most direct connections to the basic drive areas of our brains. Unlike the signals of the other senses, which first go through the brain's relay system, the thalamus, smell messages go directly to the behavior centers and are therefore least subject to rational self-control. Aromas, as a result, can bring back memories or move us to actions without our even realizing it.

In most mammals, of all the sense organs the nose was always the most important for survival. It gathered information at a distance about food, prospective mates, and danger. For the human biped, however, the eyes and ears became paramount. Nevertheless, even in today's sophisticated, supposedly deodorized, civilization, we still rely on olfaction to protect us. We can detect escaping gas in our homes and leaking fuel in our automobiles. We can smell "something burning," spoiled food, or lung-damaging pollution.

Ironically, as science has progressed to once unimaginable heights, to the moon walk, mind-control drugs, and computer technology, scientists are just beginning to realize how basic olfaction is to life on earth and how little we really know about it.

Traditionally, the sense of smell in man has been considered of minor importance in contrast to that of other mammals, reptiles, and invertebrates, whose ability to smell is critical to survival. Without it, such creatures could not locate food, identify the foe, and reproduce. While it is true that man, except for warnings against noxious fumes and poisonous food, could survive without olfaction, there is increasing evidence that our ability to smell is critical to the enjoyment of life and we use it significantly in communicating with one another in a way surprisingly similar to that of other creatures.

In 1959, researchers coined the word "pheromone" (from the Greek "pherein" meaning to carry and "horman" meaning to excite or stimulate) to describe certain substances produced by the glands of animals. Unlike hormones, the powerful products of ductless glands, which are secreted into the bloodstream to affect the animal's own development, reproduction, and behavior, pheromones are secreted externally and exert a specific effect at a distance on the behavior or physiology of another of the same species. At first, it was thought that pheromones existed as sex *attractants* only in insects. Soon scientists realized that this form of smell language was used not only in the world of insects but among reptiles, birds, fish, and mammals, including man.

There is a theory that when single cells were floating in the slime at the beginning of life on earth, it was a pheromone, or smell attractant, that caused them to aggregate to form a multicelled organism. It is now accepted that all creatures, including human beings, emit odors which affect the behavior of others.

Though we may wish to deny it, we humans are manipulated by smells just as the butterfly, the salmon, and the ape are. Our human olfactory cells are identical in construction to those of all other creatures from one end of the animal kingdom to the other. We may not have as many of them as the rabbit or the dog and we are farther away from lowly ground smells because we stand erect, but we use our sense of smell as they do.

A great deal of work is now going on in the field of olfaction. The scientists who are studying the phenomena include physicists, biochemists, sociologists, entomologists, sex therapists, physicians, psychologists, and economists. Their findings about this most magical of all our senses should give us greater appreciation and pleasure in a world in which we all smell.

2 The Supersense

SMELLY KELLY and Albert Weber both made a profession using their noses.

Smelly Kelly was a subway sniffer for the New York Transit Authority for more than thirty-four years. It was his job to sniff out gas leaks. One time, an irate tavern owner accused the subway authority of allowing fumes from the tunnels to escape into his establishment. Kelly sniffed once around the barroom to localize the odor, climbed up on a chair, tapped a spot on the tavern wall, and announced, "Dead rats!"

He was right.

Albert Weber, as of this writing, is still at work protecting us from rotten food and drink. He is the dean of two dozen organoleptic analysts—food sniffers—working for the U.S. Food and Drug Administration (FDA).

A graduate chemist with a master's degree, Weber was testing food with his test tubes and microscopes in the FDA's New York District Laboratory in Manhattan in 1943, when a call came in from the Boston office. A shipment of suspect ocean perch was on its way. There was —and is—no way to test chemically for partly decomposed fish—they have to be smelled.

Weber was elected that day and has been at it for more than three decades since. He sniffs everything from dog food to soft drinks, but he first made his reputation as an expert in fish.

The first day after he sniffed fish, the FDA chemist said he couldn't understand why people who sat next to him on the subway immediately got up and moved as far away as possible. Then his wife exclaimed when he arrived home, "My God, what were you doing today? Get those clothes off and take a shower!"

Weber smells as many as four thousand raw shrimp or five hundred fish fillets a day and rates them Class I (good commercial), Class II (slightly decomposed), or Class III (advanced decomposition). Some samples, he says, are beyond Class III and must be smelled at arm's length. That's why he does not recommend his job for other chemists looking for a good specialty.

Nevertheless, some importers are so in awe of Weber's nose that they switch their questionable shipments to ports other than New York. Early in 1975, for instance, the FDA noticed a sudden switch in a questionable fish cargo from New York to Philadelphia. Weber was sent to Philadelphia and, sure enough, confirmed that the fish stank.

By sniffing, Weber can analyze twenty-four cans of tuna in two hours. To do the analysis chemically would take a couple of days and might not be as accurate.

Another person with a highly developed sense of smell was the famous Helen Keller. Deprived of all her senses except touch and smell, she could identify friends and visitors by their personal odors. Her sense of smell was as good at helping her to recall someone's name as our senses of vision and hearing.

THE SMELL BOOK

A society matron, who wished to remain anonymous, amazed scientists in the 1940s with her ability to identify smells. She could report who last slept on a newly laundered pillowcase and was able to match coats in a closet to guests at a party by scent alone.

Such feats of olfaction are not that unusual. Most of us could equal them if we used our potential. We can, for instance, with an untrained nose, smell 0.000,000,000,000,071 ounce of skunk odor. We are able to take one sniff and identify a single aroma from among thousands we have experienced in the past. Our olfactory abilities are so keen, in fact, that we are more sensitive to changes in concentration of odors than those highly efficient smellers, the rats.

Odor memory is less influenced by the passage of time than auditory and visual memories. In one experiment, subjects were shown pictures and after a few seconds were asked to recall what they had seen. Recall was almost 100 percent. But after 120 days, it was only 50 percent. By contrast, in a similar experiment, odor recall was 70 percent immediately after exposure to scents and 70 percent 120 days later. Once remembered, smells are rarely if ever forgotten. Scientists believe this is because odors stir basic emotions. They may have no meaning themselves but they become associated with "feelings." Nothing can recall a memory as quickly and as surely as an odor. If you don't believe it, try this easy recall exercise. Imagine the odors you associate with the following:

- A garage
- A drugstore
- A coffee shop
- A dentist's office
- A spring morning
- Christmas

Memories can be instantly recalled if you catch a whiff of an odor from your childhood: the scent your mother used; your father's after-shave lotion; your home; your classroom.

How does our sense of smell work? The truth of the matter is that scientists do not know. They know how we see and hear, but they are mystified about how tiny molecules inhaled from the air can be processed and identified, and how those identifications can be filed away almost indefinitely in our memories. They are puzzled about how those odorants can trigger drives for sex, hunger, and aggression.

They do know much about the machinery involved. Take one of the most important pieces of equipment, the nose. You can't miss it. It's right there in the middle of the face, the most prominent feature. It gives our visages their character.

Symbolically the nose has always played an important part in human cultures. A Mohammedan ritual advises washing the nose with water each morning to expel the devils that supposedly visit the body through it at night. Eskimo morticians plug up the nostrils of corpses lest the soul escape from the body and become restless.

Today the nose is used to symbolize a whole range of attitudes. When we stick our noses into other people's business, we are interfering. When we stick our noses up in the air, we are snooty. If we thumb our noses at someone, we signify rejection. And, of course, if we rub noses with someone, we demonstrate affection.

But the nose's most important job is not to give information to other people, but to receive messages about the environment. In animals such as the dog the skeleton of the snout projects out well beyond the eyes. In primates, such as man, the bony skeleton of the nose is still present but very much flattened. The difference between the snout in other animals and the nose of primates is attributed to the latter's move up from the ground to the trees where most primates remain.

Because primates lived in the trees and developed a grasping hand as opposed to a clawed forefoot, sniffing became less important than good eyesight. As a result, the long snout with which other mammals explore smells on the ground was no longer necessary. As the snout shrank, the nasal barrier between the eyes decreased, allowing the fields of vision to overlap.

Thus the sense of smell, although still well developed, lost much of its importance in the apes and man, and this was reflected by anatomical changes in the brain. In addition to the general reduction in the nasal apparatus, there was a progressive increase in the size of the brain, which brought about the enlargement of the skull. This disproportionate development of the upper part of the skull produced a change in the position of the face, which shifted to the lower forefront of the head.

In mammals such as rodents and carnivores, which depend on olfaction for survival, the olfactory brain structures are relatively large and occupy all or a large part of the basal surface of the forebrain. But in monkeys, apes, and humans there is a marked reduction of all olfactory structures.

The nose itself is associated with both olfaction and respiration so that its structure and function in all creatures is an expression of that close relationship. The part of the nose which projects from the face in man is known as the external nose.

There is variety in the shape of this external nose and the underlying anatomical structures between races.

It has never been definitely proved that there is an inherent difference in olfactory ability between races, although Darwin thought that because blacks had broader nostrils and wider nasal cavities they could smell more effectively than members of the white or yellow race. Other researchers have said there appear to be anatomical differences associated with climate: the long, narrow nose which breaks up the airflow is more suitable in a cold, dry climate while the nose which allows a free stream would be more useful in a hot, moist atmosphere.

The variation between races of men is probably insignificant as far as olfactory apparatus is concerned. Unless there is a mechanical, psychological, or neurological disturbance of our olfactory system all of us can still smell very well. If such an impediment is present and we are not able to identify odors, our oxygen intake is reduced, and our emotional and cardiovascular systems may be affected.

When the nose is in good shape—literally and figuratively—it acts as a steam-heating and air-conditioning system. No matter whether the air around us is cold or warm, dusty or clean, humid or dry, the lungs receive it nicely warm, moist, and partially cleaned at the rate of about 500 cubic feet per twenty-four hours.

The conditioning system in the nose involves three small, scroll-shaped bones on either side wall called turbinates. They are arranged in layers, one above the other. Each turbinate hides a channel through which the nose is connected to those holes in our head known as the sinuses. Tear ducts empty into the lowest turbinate channel.

The turbinates humidify and warm the inhaled air by secreting close to a quart of water per day. They are helped to keep things moist by a continuous mucous membrane which lines the structures inside the nasal cavity and extends to the interior wall of the sinuses. Glands in the membrane secrete a thin blanket of mucus.

The membrane lies directly over tiny tissue "hairs" called cilia. There are six to twelve cilia to each cell in the nose. They are supple and wave back and forth 250 times a minute. For their size, their power is spectacular. In sixty seconds they perform work equivalent to lifting their own weight 14 feet in the air. Only extreme cold and injury can slow them down.

The powerful strokes of the cilia move the thin overlying blanket of mucus about half an inch per minute. The mucus from the nasal passages moves downward with the flow of gravity while the mucus from the lungs and throat moves up against gravity. In both cases, the mucus is delivered to an area in the back of the throat where we either spit it out or swallow it. We get a completely new blanket of mucus every twenty minutes.

In conjunction with the movement of the cilia, the mucus lining traps odorants, along with dust, bacteria, and other particles, and carries them away on the conveyor belt.

In the absence of a cold that blocks breathing, our noses are ever vigilant while the rest of us sleeps. Not only does the nose warn us in case of fire or other noxious fumes; it impels us to move in our sleep. Its cells congest and decongest alternately in a rhythmic cycle. When the right side becomes congested, we automatically turn over to the left side to seek more air.

One very important function of the nose is to monitor every bite of food we eat. In its position above our mouths, it checks safety through aroma. Bad food—bad odor. (Maybe nature didn't know about Limburger.) The nose also encourages us to nourish ourselves by the pleasure we derive from a delicious-smelling food. In fact, most of the sensuous pleasure we get from our meals is due to aroma, not taste, for in contrast to the myriad odors our noses can detect, we can only distinguish four tastes: salty, sour, bitter, and sweet.

Our sense of taste is powerful. We can identify quinine, for instance, in as little as one part per billion solution but taste pales next to our amazing ability to smell. We can detect an unlimited number of odors, some from far away and in dilutions as weak as one part in several billion parts of air. Arabs supposedly can smell a campfire 30 miles across a breezeless desert.

Everything has an odor to some degree but particles for either taste or smell must be soluble. This is a throwback to our ancestral life in the sea when smell and taste were one. Sugar has no taste on a dry tongue just as the scent of roses would go unnoticed in a dry nose. In order to be smelled, molecules also have to be volatile. They must leave their source and float around in the air, even if the air is still.

When you smell an after-shave lotion or the perfume someone is wearing, you smell the molecules of the scent which have drifted to your nose. The odor molecules are inhaled with the air and dissolved on the wet film of mucus in your nose, and information about the molecules is relayed by sensory cells high in each nasal passage to the olfactory bulb, where it is sent along tracts in the olfactory lobe to the brain. You then realize within a thousandth of a second that the person is wearing a particular scent. But how is the message encoded and delivered to the brain? How does the nose select one molecule over another, enabling one scent to overcome another? Why is the filing system of scent memories so efficient and apparently indestructible?

The incredibly specialized odor sensory cells, located high up in an inaccessible place at the top of each nasal passage, are pigmented yellow or brownish yellow, which distinguishes them from the ordinary cells of the nose. In man they occupy an area about the size of a dime, whereas the smell sensory area in the dog or rabbit is about the size of a handkerchief. We have about five million of these specialized cells while the German sheepdog has about two hundred and twenty million. Just as some people have a better sense of smell than others so do some dogs. The flat-nosed Pekingese and the English bulldog have a poor olfactory capacity.

The construction of these cells is similar in all mammals, including man. Each of the cells has cilia on its exposed side, and on the inside there are nerve endings that lead directly to one of the two olfactory bulbs of the brain, through the sievelike openings of the ethmoid bone located behind the bridge of the nose. The two bulbs lie on top of the ethmoid bone beneath the brain's frontal lobes.

Neuroanatomists have found the olfactory system unique because instead of going through the dorsal thalamus, where the other senses establish relay stations to the neocortex—that "new" part of the brain which gives us our intellect—the olfactory cells send their fibers directly to the brain area formerly called the rhinencephalon (from the Greek for "nose brain"). At one time this area—which is considered the oldest in evolutionary terms—was believed to deal only with smell, hence the name. More than twenty years ago, however, anatomists found that this so-called nose brain also deals with the regulation of motor activities and the primitive drives of sex, hunger, and thirst. Therefore, the term "rhinencephalon" was changed to "limbic system," derived from the limbus, or border, rimming the cortex of the brain.

Evolutionists maintain that the two cerebral hemispheres of the brain actually developed from the ancient olfactory lobes and that as the brain became more complicated the primitive nose brain remained at the forefront. The reason for its primary position, they theorize, is that olfaction was the first distant receptor which could operate efficiently in a water medium like the ocean; and, since life evolved in the sea, the first part of the brain to develop was that area concerned with smell. It is interesting that in the infant olfaction is the first sense to become dominant and to guide the movements of its entire body.

Stimulation of the olfactory bulb shoots electrical signals to an almond-shaped nugget known as the amygdala, in the area of the limbic system concerned with visceral and behavioral mechanisms, particularly those associated with sensory and sexual functions. These signals are then relayed from the amygdala to the brain stem, the "turnpike" that contains the interconnections between brain and body. Therefore, the electrical stimulation involved in smelling directly affects the digestive and sexual systems and emotional behavior. Destruction of the amygdala area has resulted in a loss of fear and rage reactions, an increase in sexual activity, excessive eating, and severe deficiencies in memory.

It has been observed that when epileptics have a seizure caused by a trouble spot in the temporal lobe, a part of the limbic system, they smell strange odors just before the attack. When areas on the temporal lobe are stimulated electrically by a researcher, conscious patients will report odor sensations.

In 1937, Japanese researchers first reported there was electrical activity in the nose's sensory cells and in the brain when an odor stimulus occurred, but it was not until the 1940s and 1950s that such electrical impulses were measured systematically as to strength, duration, and quality. The brain's electrical response to an odor—about 40 cycles per second—appears to be indistinguishable from that correlated with emotional behavior. Since olfactory signals are sent to the amygdala area it is easy to explain how what we smell affects our emotions and our sex and hunger drives.

There are many conflicting opinions among researchers as to how the brain identifies smells. Some maintain that the olfactory bulb has been mapped for specific odors. For example, the inhalation of fruity odors activates the front part of the bulb, while solvents, such as benzene, stimulate preferentially the back part of the bulb. Until recently, the scientists did not know that the trigeminal nerve in the cheek is involved in mediating certain chemical sensitivities to smells. The trigeminal receptors, which are bare nerve endings dispersed in the nasal passages, mouth, and throat and in mucosa around the eyes, communicate with the brain via the cheek's trigeminal nerve. In principle, an odor can be sensed through the interaction with either these receptors or the smell sensory cells or with both. The degree of participation of either depends upon the nature of the odorant and its concentration in the air. It is thought, but so far not too well documented, that harsh odors such as ammonia provoke considerable trigeminal responses whereas lighter and more pleasant odors are sensed primarily by the olfactory pathways.

A unique and as yet unexplained phenomenon of smell is that of adaptation. You know that no matter how strong an odor is when you enter a room you become unaware of or "blind" to it a few minutes later. When subjects are tested for recognition of certain odors, there has to be a pause of at least thirty to sixty seconds between stimulations before any odor, new or the same, can be recognized. It has also been shown that prior exposure to one odorant will decrease the sensitivity to another. A higher concentration of the second odor is therefore required for detection.

When scientists record the electrical activity that smelling an odor causes in the olfactory bulb, they find that even though the subjects can no longer detect the odor—say, of roses—the electrical signals continue unabated in their brains. In other words, the physical stimulus—the rose odor—still is at work, but the subjects are no longer conscious of the scent. A recent study has shown that animals exposed to the same odor over a prolonged period of time show degeneration of the nerve fibers carrying smell messages to the olfactory bulb in the brain.

Why is there adaptation? No one knows for certain, but some theorize that it is somehow involved in the protection of respiration. Perhaps we would stop breathing, or be unable to concentrate, if we were constantly aware of an odor, or our other senses would be overpowered by our concentration on odor.

Adaptation is just one of the many mysteries about olfaction. There have been at least sixty theories of how the brain receives and interprets information from the nose. Most can be categorized as chemical theories. These hold that molecules or particles of odorants touching the olfactory cilia are absorbed, creating an electrochemical change in the nose's sensory cells, which then send electrical signals to the brain. Some investigators believe that enzymes, the body's catalysts, are somehow involved in the recognition and relay of odor information. Others theorize that specific odor molecules fit into specific receptors in the nose just as a round peg fits into a round hole. Still others, prominent among them Dr. John Amoore of the U.S. Department of Agriculture, who works at their Western Regional Research Laboratory, feel that there are primary odors much as there are primary colors and that the nose recognizes combinations of these primary smells.

Dr. Amoore believes that there are at least thirty primary odors and that, just as we combine primary colors such as blue and yellow to make green, we combine the primary odors to create the myriad of smells in our environment. He claims to have succeeded in isolating four primary odors, three of which (isovaleric acid, 1-pyrroline, and trimethylamine) are suspected of being primate, and even human, pheromones. The fourth (isobutyraldehyde) occurs in a wide variety of foods, and its malty odor may signal the presence of three indispensable amino acids needed in our daily diet, since we do not manufacture them in our own bodies. Dr. Amoore says that the primary odors yet to be discovered may provide sensory input about foods, localities, and predators, but the most intriguing are probably the pheromones. Do we humans respond to sexual scents just as other creatures do?

3 The Scent of Sex

UNTIL VERY RECENTLY, it was believed that chemical smells which affect the sexual behavior of all other forms of animal life have no part to play in human social and sexual intercourse. Now more and more evidence is accumulating that we are profoundly influenced both physiologically and psychologically by the volatile secretions of others. Sexual attraction, which we have intellectualized and romanticized, may really be a response to an invisible, silent stimulus—a primal animal scent.

This is not a new concept. Somerset Maugham, curious to discover the secret of H. G. Wells's success with women, reported, "He was fat and homely. I once asked one of his mistresses what attracted her to him. I expected her to say his acute mind and sense of fun; not at all; she said that his body smelt of honey."

Humans have, of course, since ancient times borrowed animal sexual scent products such as musk from the musk-ox, civet from the civet cat, and castor from the beaver. The influences of human smells on human sexual behavior were, however, all but ignored by scientists until 1871, when Charles Darwin stimulated interest in olfaction and sex with his controversial book *The Descent of Man*. Although he described smell as a powerful sex attractant among mammals and insects, he maintained it was not as important as it probably once was among humans. He did admit that the power of smell differs greatly among individuals.

Then in 1891 Ernst Haeckel, a German embryologist who believed that the human fetus passes through all the stages of evolution from fish to reptiles to birds and finally to mammals, asserted that in the beginning, when single cells floated around in the slime, there had to be an olfactory attraction—a chemotropic interaction that operated through water—to bring cells together. This rudimentary olfactory attraction continued to exist, Haeckel argued, and remained operative in human sexual stimulation and attraction. Smell sensuality was a vestige from the ancient past, when it alone brought cells together. Haeckel's theory was widely accepted at the turn of the century but faded as civilization became more and more obsessed with deodorization.

In the late 1890s, Sigmund Freud's colleague, Wilhelm Fliess, spent a great deal of effort trying to prove the connection between nasal and sexual processes. He observed that during menstruation the capillaries of the nose swell and that the application of cocaine to special "genital spots," which he identified in the nose, reduced menstrual pain. This led Fliess to conclude that there was a periodicity of human behavior. He was not concerned so much with olfaction as he was with tracing the evolution of the nose from its use by lower animals in finding sexual partners to its sexual functions in humans.

A decade before Fliess, the German biologist Gustav Jaeger was already convinced that smell plays an essential role in human sexual intercourse. In a series of essays published in 1881, he went so far as to say that odor was the "origin of the soul."

A Frenchman, Auguste Galopin, elaborated upon Jaeger's theory in 1886 and also proclaimed odor as a central factor in human love. In *Le Parfum de la Femme*, he said the mutual interaction of odors constitutes the essence of sexual love: "The purest marriage that can be contracted between a man and a woman is that engendered by olfaction and sanctioned by common assimilation in the brain of animated molecules due to the secretion and evaporation of the two bodies in contact and sympathy."

Fliess, Jaeger, and Galopin insisted that the sniffing so long associated with the sexual life of animals was essential to human love life as well.

The fact that smell plays a part in the human communication known as "kissing" was recognized in the Bible. When blind Isaac asked Jacob to kiss him before bestowing the divine blessing meant for Esau, Isaac's intention was to identify his son by smell. But Isaac's nose was fooled by Jacob's wearing of Esau's clothes.

The nose is directly over the mouth, so that what we kiss we also smell. Many people—the Eskimo, Maori, Samoans, and Philippine Islanders, for instance—are direct about it. They rub noses or place their mouth and nose against the cheek of another person and inhale as a means of identification. In some of their languages the word for kissing means "smelling." And Borneans never "greet" anyone; they "smell" them.

The Arabs breathe in each other's faces while talking because they think it is an insult to deny another one's breath. It is also an Arabian custom for a man's relatives to smell a girl before they select her for his bride.

As with other creatures on earth, odor serves mankind as an indication for acceptance or avoidance. As well as our sexual odor each of us has racial, cultural, and family smells which help others to identify us. Whether we are accepted by individuals, groups, or a sexual partner depends a great deal upon a combination of these scents.

The upper classes, of course, have long thought that the lower classes smell, and in a way, that is true because soap, hot running water, and dry cleaning may not be readily available to the poor.

The middle and upper classes in our society support their status by the use of deodorants and re-odorants—expensive perfumes. In fact, perfume advertisements in middle- and upper-class magazines today often show the trappings of an aristocratic life including furs, jewels, expensive cars, and representations of royalty.

Prostitutes through the ages, aware of their "unclean" status in society, have tried to cover up their "moral malodor" with perfumes and, indeed, an insult directed at an overly perfumed woman may be that she "smells like a whore."

What we eat and how we live do affect how we smell and how strange others smell to us.

Minority groups have often been stigmatized in terms of odor—garlic or curry, for instance. To hide such scents, some have resorted to strong-smelling perfumes. Central Europeans carry the scent of the cabbage, turnips, and radishes they consume. Indians smell of the rice and spices in their diet, while the South Sea Islanders have a body fragrance of fruit and palm. Americans smell like butter to the Japanese, who smell like fish to Americans.

To the rest of the world, the Eskimo stink of blubber, oil, and sweat, and yet they have a more acute sense of smell than most people on earth. They develop it as infants carried in their mother's parkas, and as they grow up naked in warm igloos heated by burning animal fat.

Actually, Eskimo, Caucasians, Africans, and Orientals all give off different body odors, regardless of what they eat and how they live, because they have different distributions of the specialized scent glands called "apocrines."

Mammals send messages with apocrines, which are modified sebaceous glands. Ordinary sebaceous glands emit a fatty substance to keep hair and skin lubricated, while apocrines send out scent signals. Humans, in addition to sebaceous and apocrine glands, also have eccrine (sweat) glands. The eccrines constantly secrete small amounts of perspiration and are located over almost the entire body.

Eccrine sweat tastes and looks like salty water. Its function is to help the body dissipate excessive internal heat. The skin and its sweating mechanisms are vital parts of the body's temperature-regulating system. A thermostatlike center in the brain sends out flashing nerve orders when the body gets too hot. The heart pumps more blood to the skin so that heat is carried off. A little heat is radiated from the skin but the most important means of body cooling is evaporation of sweat. This process goes on even though the skin may seem dry. Sweat glands put out two or more quarts a day under some extreme circumstances.

The eccrines on most of the body respond only to physical stimulation, such as exercise and heat, with the exception of cases of extreme emotional arousal when the whole body may break out into a "cold sweat," so called because at these times the body is not heated up and the evaporation of moisture causes a chill. The eccrine glands crowded on the palms of the hands and the soles of the feet, however, increase their output during fear, pain, tension, and sexual excitement. Those located under the arms respond to both heat and the emotions.

In areas of the body where eccrine sweat cannot evaporate easily, such as the underarms and the feet, the sweat can become noticeable and uncomfortable, especially when the humidity is high. Metabolism also affects the amount of sweating. But, contrary to popular belief, perspiration has no odor. It is the action of bacteria that produces the unpleasant odor of stale sweat.

The amount of eccrine secretion and odor is approximately the same for both sexes but those who do not shave under their arms tend to have a stronger odor because hair acts as a collection ground for the decomposition products of bacteria. (Incidentally, it has also been determined that you sweat more under the arm you use most so that if you are right-handed, you might need more antiperspirant under your right arm.)

Hair under the arms and around the genitals is specifically designed to collect the odor of the secretions of the apocrine scent glands. As in all other mammals, human apocrines are small until puberty, when they are stimulated by hormones. Hair under the arms and around the genitals is also sparse until puberty. The apocrines secrete their substance through the narrow pit of the hair follicles, just as the other sebaceous glands secrete sebum. Like sebum, it is somewhat oily compared to the watery sweat of the eccrines. But while sebum is used to keep the skin protected and supple, the as yet unidentified chemicals emanating from the apocrines have another purpose. They send out the scent signals we call "body odor."

In a recent study of sexual odor by research psychologist Michael J. Russell at the University of California Medical Center in San Francisco, freshman college students described the male odor as "musky" and the female odor as "sweet." The experiment involved sixteen male and thirteen female students who had not used soap, perfume, or deodorants for twenty-four hours before the test. During that period they all wore white T-shirts as undergarments. When the shirts were collected for the test, each was placed in a cardboard ice bucket with a hole in the top so that the underarm portion of the shirt was nearest the hole.

Initially, each subject was given three buckets to sniff: one containing his own shirt, the second the shirt of another member of the same sex, and the third a garment worn by a member of the opposite sex. Some minutes later, the students were offered a choice between two buckets, one with a male shirt and one with a female shirt. (The bucket with their own shirt had been removed.)

In the test thirteen out of the sixteen males and nine out of the thirteen females correctly identified which shirt had been worn by a male and which by a female. Some said afterward that the distinction was obvious and that a lower level of odor would have been adequate. None said they found the odors objectionable.

To determine whether the ability to detect male or female odor depends on puberty, Russell then repeated the experiment with a group of nine-year-olds. The youngsters had an equally good score in determining which shirts belonged to males and which to females. He intends to repeat the experiment with two-year-olds, as he firmly believes that being able to discriminate sexual odor is innate in humans.

Russell points out that the human male is smellier than the female, something which is not true in most other primates, and assumes that a human pheromone does not necessarily have to have a pleasant smell to be an attractant.

The principal odor-producing parts of the human body are the glands of the genitoanal region and under the arm. All humans have apocrine glands around the anus. The hairiness and warmth of the human underarm and genital region are ideal for the diffusion of scent. In Russell's opinion, however, the precopulatory odor cue is not localized: "I believe it is a whole body scent, although I do think the face, which is loaded with apocrine glands, is important. That's why humans like to kiss."

There are certain regions of the body where the apocrine glands are present in some races but not in others—again reinforcing the human use of scent for identification.

In Caucasians, particularly Europeans, the apocrine glands under the arm are so numerous and close together that they almost form one rounded object. Unlike the eccrine glands, which look like corkscrews, the apocrines are straight tubules which feed directly to hair follicles.

Blacks and whites have numerous apocrines under the arm, in the anal and genital region, and on the chest and around the nipples as well. Blacks have more than whites. From the racial point of view, the most striking difference in the distribution of apocrines is their weak development in Orientals. Orientals, of course, have little body hair. The Japanese are somewhat of a mixture of races, so they have more apocrine glands than the Chinese but far less than whites and blacks. The Koreans are almost devoid of apocrines and have little body odor even when careless about hygiene.

A remarkable correlation has been found between the degree of development of the underarm apocrines and the type of earwax produced. In blacks and whites, who have well-developed underarm apocrines, the earwax is generally soft and sticky, although dry earwax does occur in some Europeans. In the Oriental races, in which the underarm organs are scarcely or not at all developed, the earwax is dry. The smell of the sticky ear substances is believed to be a repellant to insects which might wander into the ear canal.

The Japanese have almost no apocrines around the genitals, although they do have them around the anus, and they consider underarm odor a sickness. In fact, it is considered a medical reason for being excused from service with the armed forces. There are even doctors who specialize in treating it, and afflicted patients are admitted to the hospital.

Even within races and subraces there are different body odors. The natives of Angola, for instance, are said to have a particularly strong body odor while those of Senegal are less odiferous. Some Europeans claim to be able to distinguish between African tribes by scent alone, and Europeans of Nordic descent are said to have a stronger smell than other persons of the same race. Peruvian Indians are reported to be able to distinguish the odors of Europeans, blacks, and their own and have special words for the smells. Blacks living in the Antilles are reputed to be able to distinguish between the smells left behind by a Frenchman and a black.

John Baker, emeritus reader in cytology at Oxford University, in his book *Race,* published in 1974, concluded that apocrine glands probably help to identify individuals as belonging or not belonging to a group. He noted, however, that there are many other factors which contribute to an individual's odor. Among them are customs, food decaying between or in teeth, the smell of the breath, whether opium or tobacco is smoked, and objects adhering to the body or the clothes. Beyond these are innumerable substances deliberately applied to the body for the sake of their odor.

There is no doubt that animals communicate with

a variety of scents, including those emanating from urine and feces. But their strongest messages are transmitted by the products of their apocrines.

Since all humans have these special scent glands to varying degrees, it can be assumed they communicate with them. The fact that the apocrines are present in humans at birth but do not begin to function until puberty, when the pubic hair appears to trap the apocrine scent, seems to indicate that they are important to sexual relations. Their peak size is reached at sexual maturity and diminishes with age. Hence children do not have the characteristic body odor of adults and elderly people have less body odor than younger adults, lending weight to the belief that apocrines in humans are sex-pheromone producers.

One of the primary odors that Dr. Amoore claims to have isolated in the course of his work at the Western Regional Research Laboratory is trimethylamine. The odor of trimethylamine is well known to organic chemists, who describe it as "fishy." It is pronounced on dead fish which are not refrigerated and is also formed by bacterial action on betaine, sometimes tainting the milk of cows fed beet tops.

According to Dr. Amoore there is a good deal of indirect evidence that trimethylamine may be an important mammalian sex attractant, a human pheromone. The Swedish botanist Carolus Linnaeus noted in 1756 that the domestic dog is extremely interested in the odor of the plant called stinking goosefoot. Linnaeus named the plant *Chenopodium vulvaria* for obvious reasons—it smells like human menstrual blood. Its tissue contains a large amount of trimethylamine.

Trimethylamine is prominent in human menstrual blood and it is a quite well-known phenomenon that the odor of menstruating women brings many male animals into a state of sexual excitation. This suggests that trimethylamine might be a common estrus-signaling pheromone for several mammalian species.

Dr. Amoore said the value to the human female of trimethylamine as a sex attractant remains rather obscure, since ovulation occurs some ten days after menstruation ends.

Most researchers are still reluctant to state that human sexual pheromones definitely exist, although pheromones have been identified in other mammals. When the female mouse, for instance, smells the product of the apocrine gland located on the male's penis, she is strongly attracted to him, even though she is not in heat. The male then urinates and another pheromone in his urine apparently brings her into heat. In other words, the preputial pheromone attracts the female mouse and the urine pheromone makes her sexually receptive.

An intriguing aspect of this is that virgin female mice not under the influence of estrogen show little attraction to the male penis pheromone, but a sexually experienced female, except when pregnant, finds it irresistible whether or not she is in heat.

The mouse penis pheromone has the opposite effect on strange males of the same species. Instead of intriguing them, the secretion infuriates them. The male urine pheromone serves also as a warning to subordinate males to stay away from any territory sprayed.

The coordination of hormonal output between the sexes is obviously important for successful reproduction, which in the females of all mammals, including humans, depends upon the attainment of sexual maturity, rhythmic functioning of the adult ovary, and the maintenance of pregnancy. It is known that in the female mouse each of these events can be modified by social stimuli. The presence of an adult male during the early growth period of the young female mouse accelerates the onset of sexual maturity. Similarly, the presence of an adult male induces a synchrony of ovarian function in the adult female. Exposure to a strange male during the early stages of pregnancy can cause the female mouse to lose her offspring.

W. K. Whitten of the Jackson Laboratory, Bar Harbor, Maine, first reported in 1956 that the female mouse's estrus cycle was affected by the presence of the male mouse. Moreover, he found that when he treated the nose of a female with local anesthetic, so that she could not smell, her estrus cycle was changed. He said female mice, in order to maintain the hormonal cycle, must be able to smell a pheromone in male urine.

Whitten also suggested that a pregnant female mouse may lose her fetuses when exposed to a strange male because of the action of the stranger's pheromone on her pituitary. The pregnant mice that lost their fetuses did have low levels of pituitary hormone soon after exposure to the strange male.

The influence of the male mouse on the female reproductive system has been shown to be dependent upon a volatile substance found in male urine. This substance includes testosterone, thus indicating that the hormone level of the male influences the sexual development and reproductive behavior of the female.

Is there a similar link between human sexual behavior and the products of scent glands?

In the late 1800s, a phenomenon called the French Boarding House Syndrome was reported in medical literature. Girls living in boarding houses and isolated from men entered puberty later than girls exposed to males even when the latter had no personal physical contact with men. Somehow the proximity of the male sex affected the females' hormones.

Some unknown substance also affects the hormones of girls who live together for any length of time. Both British and American researchers reported in the early 1970s that when young women lived together in a college dormitory their menstrual cycles tended to synchronize. When these same women were exposed to males more than three times a week, they had normal cycles of twenty-eight days. The girls who saw less of males had longer cycles.

The mysterious influence of the proximity of the opposite sex appears to work in both directions. In his book *Lives of a Cell* Dr. Lewis Thomas mentions a scientist who lived for long periods of time in isolation on an island. He discovered, by taking the dry weight of the hairs trapped by his electric razor every day, that his beard grew much more rapidly each time he returned to the mainland and associated with females.

One of the first modern reports linking human sexual behavior and olfaction was made in 1952 by the French researcher J. Le Magnen. He states that for normal women olfactory sensitivity to musk chemicals varies significantly during the menstrual cycle, reaching its peak at the time of ovulation, when it is 100 to 100,000 times greater than at menstruation. Women who had had their ovaries removed were found to be far less sensitive to musk, which is closely related chemically to the male hormone testosterone, but the majority of those examined retained normal acuity when treated with estrogen. On the other hand, adult males and prepubescent youngsters of both sexes were almost completely odor blind to musklike smells.

Although not all researchers have been able to reproduce Le Magnen's work, Dutch and American investigators have reported that there are two peaks of a woman's ability to smell—particularly sexual scents—during each menstrual cycle, one preceding the ovulatory stage and one during the luteal phase, eight days before the menses. Furthermore, an American clinician, Dr. Robert Henkin of Georgetown University Medical School, has reported that women who do not menstruate and who have a poor sense of smell are less likely to recover their fertility than those who do not menstruate but who retain their ability to smell.

Researchers have found that after menopause, natural or surgically induced, women not receiving estrogen replacement often had poor olfactory acuity. Furthermore, testosterone, the male hormone, when given medicinally, worsened a woman's ability to smell. In still another study, it was discovered that sensitivity toward musk decreases during the first months of pregnancy and increases up to the time of delivery. Thus, there is increasing evidence that hormone production affects a human's ability to smell just as smells affect our ability to produce hormones.

Sex pheromones are produced by both sexes. Among the pheromones identified so far among mammals, the male sex pheromones seem mainly to function as aphrodisiacs for the female while the female sex pheromones apparently announce her sexual readiness.

Some researchers are convinced this is true in humans. In fact a group of English researchers maintain that the sweat and vaginal glands of women secrete odors in response to stimulation by the hormones that are most plentiful in their bloodstream just when their menstrual cycle is in the phase of maximum fertility. Men, on the other hand, produce a scent which increases in intensity with the degree of their sexual arousal. This, they assert, would explain why women may be subliminally stimulated by an aroused male within a crowd, say, at a cocktail party. The pheromones unconsciously secreted by the man send a message to the woman who is, herself, emitting a silent, unconscious message of willingness and readiness.

Such a dialogue has been proved in animals, whose sense of smell is vital to the survival of the species. Nature endows them with the olfactory ability to choose a mate very selectively. For instance, a female rat prefers the scent of a normal male over one which has been castrated. She also prefers to mate with a male whose odor does not show that he has copulated recently with another female. The males are more attracted by the odor of a new female than by that of one with which they have just copulated. This is an olfactory method of ensuring that several females will be impregnated and the next generation will be assured.

There may be debate about human susceptibility to pheromones, but external chemical messengers—odors—have been shown to penetrate the human subconscious. Within seconds after exposure to an unnoticed olfactory stimulus the electrical resistance of the person's skin decreases and changes occur in the blood pressure, respiration, and pulse rate. It is assumed that a volatile chemical —not necessarily detected as an odor—causes changes in the brain.

Smell messages have been traced to the region of the human brain linked to hormonal control of reproductive functions and sexual behavior. For instance, in a congenital condition called Kallman's Syndrome, involving the absence of the olfactory bulbs and a defect in the area of the brain known as the hypothalamus, the releasing factors—secretions which control the release of pituitary gland hormones—are affected. This, in turn, results in very low levels of sex hormones circulating in the blood and in underdeveloped sex organs. Since the syndrome involves both the olfactory system and the area of the brain concerned with the release of sex hormones, a relationship between the sense of smell and human reproductive physiology would seem highly likely.

In rats that have been deprived of their sense of smell at birth the levels of growth hormone produced by the pituitary gland are low, they are stunted, and their testicles are subnormal.

There is more to sex than just pheromones, of course. Even in rats, it has been demonstrated that when their olfactory bulbs are destroyed experienced male rats will continue to copulate, although their ejaculations will be decreased. Virgin males, however, will not pursue amorous activities of any sort.

When female rats with their olfactory bulbs destroyed are given testosterone, they mount receptive females as frequently as do experienced males with olfactory impairment. Since such homosexual mounting is dependent upon the condition of the stimulus female, researchers believe that, since the impaired rats cannot smell, they may be responding to body language as well as to their own internal hormonal condition.

Olfaction may not be the only factor in sexual behavior, but it is apparently very important to our cousins, the monkeys and the apes. Most nonhuman primates will mate only, or mainly, during the period of the sexual cycle when the female is in heat. Many primate males are strongly attracted to the scent marks of females in estrus and also sniff and lick the genitals of these females, thus gathering information about the condition of their prospective sexual partners. Female spider monkeys show no obvious visual signs of being in heat, unlike most other nonhuman primates, and the males of this species sniff and lick the females, and even drink their urine, in all phases of the sexual cycle. Presumably, in default of visual clues, they depend on smell as the main transmitter of sexual information.

When the females of some primates are treated with the female hormone estrogen, they produce a pheromone which arouses the sexual interest of the males, even though the females are not in estrus. Experiments with free-ranging rhesus monkeys have shown that when the females are implanted with estrogen pellets, they become sexually receptive and the males, in turn, show not only sexual interest but also a hormonal readiness to copulate at any time of the year, although normally these monkeys breed seasonally.

The fact that information about sexual readiness is communicated largely by smell is borne out by a study of a troop of bonnet monkeys conducted by Indian researchers. There were stray mountings throughout the study period, but the vast majority occurred during the heat periods of the females. One of the males in the troop was diseased almost to the point of blindness and completely impotent. In spite of this he would, when in the vicinity of an estrous female, be drawn to her, test her, and eat her vaginal discharge. If this male was aware of the female's condition in spite of his semiblindness, it could only have been through olfactory signals, and it is interesting that his impotence did not diminish the attraction of the female in heat.

Chimpanzees, it has been observed, mate throughout the sexual cycle of the female, although most copulations take place during maximal genital skin swelling in the female, the time she is most fertile. Researchers have concluded that the female sexual cycle is relatively unimportant in the behavior of these sexy apes because social factors and individual choice override the telltale signs of "heat." They feel that chimpanzees, like their human cousins, exhibit a degree of independence from hormonal control commensurate with their relatively advanced capacity.

Curiously enough, gorillas, which are also advanced from the standpoint of evolution, seem to be largely restricted in their sex lives to the period of from one to four days when the female is in heat. At this time the females who, contrary to the normal primate pattern, initiate sex, more frequently invite the males to copulate and the males are more likely to accept the invitation. Copulations during this period tend to result in more ejaculations than at other times in the cycle, and oral-genital sex is also practiced at this time. This suggests to researchers that hormones rigidly control the sexual behavior of these big beasts.

Volatile acid chemicals secreted by sexually ready female monkeys were isolated by Dr. Richard Michael, now at Emory University in Georgia, who named them "copulins." He then proceeded to isolate almost identical compounds from human vaginal secretions. In one study Dr. Michael analyzed secretions in the vaginal tampons of fifty women volunteers during an entire menstrual cycle. He and his co-workers were able to link peak production of the vaginal secretion acids to the specific times of ovulation in the volunteers.

Acid levels were significantly lower in women who were taking the birth-control pill than in those who did not, and their acid production rate was essentially constant, independent of the menstrual rhythm. This suggests that birth-control pills in some way interfere with the growth of normal vaginal flora responsible for acid production.

Dr. Michael concludes from his work that the same volatile fatty acids acting as sex-attractant pheromones in infrahuman primates also occur as normal constituents of the vaginal secretions of young women. He believes that the data suggest a significant role for olfaction in human sexual behavior, although its importance clearly varies considerably from couple to couple.

Dr. George Preti and Dr. Richard Doty of the Monell Chemical Senses Center of the University of Pennsylvania, in another study of vaginal secretions, have discovered that an adult human female's vaginal odor varies during the month. They submitted scents extracted from secretions taken from women during their entire menstrual cycles to volunteers who did not know the origin of the samples. The volunteer sniffers selected extracts from the women's fertile period as being the most pleasant.

Dr. Preti says that secretions from the vagina are a fairly complex mixture of organic compounds, the type and the amount of which differ from subject to subject. He pointed out that a large number, possibly the majority, of the aromatic compounds in vaginal secretions are derived from the action of bacteria on the fatty acids, just as the action of bacteria on the skin is responsible for the odor associated with sweat.

In the meantime researchers at the Department of Maternal and Child Health at the University of North Carolina are conducting tests with copulins, the so-called "Michael's Mixture." In an unpublished study of sixty-two couples, a subgroup of twelve couples whose patterns of sexual behavior were cyclic showed increased sexual intercourse following exposure to the copulin mixture as compared with exposure to control substances. The North Carolina researchers concluded that their data did not reject the hypothesis that an olfactory cue influenced the desire of some human males for coitus. The couples studied, however, were not typical, as they were mostly unmarried, in their early twenties, without known gynecological abnormality, and in the upper middle class. The researchers are now studying a larger population.

While the apocrine glands may well contribute the most to the aroma of sex in humans, urine probably also emits silent signals. Human waste water contains at least sixty volatile ingredients.

Animals converse with urine. It plays a very important part in determining social distance. Rabbits of both sexes use it in the course of aggression as well as mating behavior. As precopulatory behavior, the male rabbit urinates on the female, which thus carries the male's personal olfactory signal demonstrating ownership, since his urine contains male pheromones. When a human male gives a woman perfume containing the male pheromone musk, he is doing the same thing: marking her with male scent just as male animals of lower species mark their mates to show ownership.

Male rabbits and dogs press their tails down tightly when they are frightened to suppress dissemination of their odor. In this way, they are less irritating to other, superior, males. While urine and/or penis gland odor from strange males elicit attack behavior in males, these odors are attractive to the females of the same species. Female urine, on the other hand, contains a substance which lessens the probability of attack behavior by strange males. Thus females can move relatively unimpeded through male territory. This allows the females a wider choice for a mating partner.

Odor plays an intriguing role in nature's selection of the fittest. The size and output of the smell glands in animals correlates with their status position in the group. These glands, which are generally small or absent in females, vary in size and output in relation to a male's rank. They are hormone dependent, and males generally have less glandular output during nonreproductive seasons.

In animal studies, whether or not the male is dominant, it has been demonstrated that the female most often selects her mate. Females of many insect species exercise careful control in the determination of mating. Female fireflies, for example, use the most distinctive features of male flashes to choose the correct partner. A fruit fly will approach his ladylove and display his wing vibrations. If she has already mated, belongs to a different species, or is too immature, she rejects him and produces a repulsion signal.

A female fish seeks out a male in his guarded territory. She signals her willingness to mate to ward off an attack and to check his hostility. He responds by showing his most attractive colors or swimming a courtship ballet.

Female birds of some species select the most attractive partners. Darwin noted the fact that females, by naturally selecting the strongest and most beautiful males, assured survival of the fittest.

If it is the female who selects the mate, then it makes sense that the female has a better sense of smell. The superior sense of smell in females, which has been proved in many animal species and alleged in humans, is probably necessary for survival. Female discrimination most likely prevents cross-mating between species, assuring procreation. Theoretically, males can increase their offspring by mating with more than one female whereas females cannot increase theirs by mating with numerous males. Errors in mating selection thus are far more serious for females than for males. In some species, a mating which results in sterile or inviable offspring might claim the entire season's production for the female, while the male would lose no more than a few seconds and some readily replaceable sperm cells.

The superior olfaction of the female also plays a role in two other crucial functions of the sense of smell: the establishment of sexual identity and the maternal response to offspring. It has been demonstrated that early odor experience influences the discriminative behaviors of young females more markedly than those of males. Female mice reared only by their mothers in the absence of adult males show in adulthood no sexual preferences and are indifferent to male odors, suggesting that young females are normally imprinted by the odor of their fathers. When lactating rat mothers are injected with a smelly substance, citral, which then contaminates their milk, the female offspring grow to prefer the citral odor in mother's milk more than male pups.

The odor of the mother rat inhibits the activity of her infants. It keeps them near her and out of harm's way. The mothers recognize their own offspring by smell. If a mother's olfactory bulbs are destroyed, she may still instinctively retrieve a wandering pup, but it will not necessarily be her own. She may not recognize her own pups and may eat them. This has led researchers to believe that there is a "family odor" and that a mother can recognize her own offspring from among others while her offspring can select her from among the crowd.

Recent experiments by Michael Russell at the University of California Medical Center have demonstrated that by the age of six weeks human nursing infants can distinguish odors associated with their mothers from those of other women.

In his tests, nursing mothers were asked to wear a cotton sponge inside their brassieres for several hours. Then three sponges were successively held near a sleeping baby's nose: one that had been worn by its mother, a second one worn by another mother, and a third that had not been worn at all. If the baby responded to a sponge, it was allowed to doze off before the next one was presented.

At two days of age, only one of the ten babies in the experiment responded, turning its head or sucking when presented with two sponges worn by its own mother and by another mother. But test results showed that when the babies were six weeks old, their olfactory response was more fully developed. In this experiment, Russell added a sponge with the odor of cow's milk. Seven of the ten babies responded only to their own mother's sponge. One of the seven also was aroused by the strange mother's sponge and by the cow's milk odor, responding to them negatively, however, with a head jerk and a cry.

Is there a family odor among humans? We have seen that there is often a group odor for various cultures and races. There is much yet to learn, but one thing is obvious—our denial of our ability to smell affects our sexual pleasure. It inhibits our practice of oral-genital sex, for instance, although such interaction is common throughout the animal kingdom as an expression of sexual information and pleasure.

Desmond Morris, in his book *The Naked Ape*, theorized that civilization has repressed olfactory sex stimuli because of the constant intermingling of strangers in large groups. But, he points out, the female who so assiduously washes off her own biological scent proceeds to replace it with perfumes, which are, in reality, no more than diluted products of the sex glands of animals.

The anxiety that we humans have been made to feel about our own odors—particularly our genital odors—may well be misplaced and sexually destructive.

4 Sickly Smells

A PARAGUAYAN HEALER once built up a flourishing medical practice by making long-distance diagnoses from patients' shirts and underclothing. An American who sent Tupa Mbae his socks was told by the diagnostic sniffer that his ailment was hookworm. The Paraguayan's diagnosis was confirmed by the American's own physician.

The idea of diagnosis by smell may seem ludicrous in this day and age of multiphasic chemical testing and sophisticated electronic devices. The fact remains, however, that just as each of us has his own unique odor, each disease gives off its particular telltale scent. That scent may give a more rapid diagnosis than any medical device. In fact scientists are only now working on instruments which they hope will equal or perhaps surpass the nose in odor detection.

We have all been exposed at one time or another to bad breath, smelly feet, and stinking armpits. The changes in human body odor can be due to mood, diet, or living habits. Our skin, hair, and clothing are great absorbers of scent molecules. But body odor can also be due to disease.

A particular illness mixes its own scent much as

a perfumer creates a fragrance. The ingredients consist chiefly of the secreta and excreta of the body—sweat, sebum, mucus from the nose, throat, and lungs, urine, stools, vaginal products, as well as wound discharges and decomposing tissues. Such smells are stigmatized by our culture as disagreeable. According to Dr. Ralph Crawshaw, a Portland, Oregon, psychiatrist, this quirk of culture sets up a major barrier between patient and physician.

Dr. Crawshaw, writing in the medical journal *Prism*, calls this barrier the "bedpan factor" and maintains that medical practitioners must learn to deal with excreta and fight the impulse to withdraw from a fellow human who has become "dirty."

"If this fight has been won, the aide, nurse, medical student, or whoever it may be, has learned a great lesson in humaneness, removing a barrier between himself and a sick human being . . . be it a suppurating fistula in a terminal tuberculosis patient or the snotty nose of a malnourished child, the patient is unwittingly testing and dividing medical personnel into two great classes of servants of the sick—those who have chosen to carry bedpans and those who have not. That is the bedpan factor."

Indeed, in the less deodorized and sanitized cultures of the not too distant past doctors were well aware of the diagnostic significance of smells. They didn't have the sophisticated technological devices we have today. They had only their eyes, ears, and noses. Their noses often proved to be their most precise tool. In the early 1930s, for instance, a famous New York physician was implored to make a house call on an eminent citizen of Newark, New Jersey, whose illness had defied diagnosis. The doctor was smoking a cigar at the time he was admitted to the elegant home. He took the stogie out of his mouth, sniffed, and informed the patient's wife that there would be no need for him to go upstairs and examine her husband. "Cancer!" the doctor announced, jamming the cigar back in his mouth. And indeed, a few months later, the Newark man died of that disease.

Certain types of cancer have a fetid smell. Sigmund Freud, when dying of cancer of the jaw, was terribly distressed that his favorite dog, a chow, avoided him because of the foul odor from the malignancy.

Among the other diseases that were once commonly diagnosed by smell alone were yellow fever, which had a "butcher shop" odor; scurvy, which smelled putrid, as did smallpox; and typhoid fever, which made patients smell like freshly baked bread. Diphtheria victims produced a sickeningly sweet scent, and plague patients had the odor of apples; measles victims still are scented like freshly plucked feathers, and a number of skin disease patients—those suffering from eczema and impetigo—smell moldy.

No one has to go to a physician even today to diagnose the victims of bromidrosis, a disorder of the sweat glands. In this disease the sweat smells very bad for some as yet not clearly identified reason, although the cause is believed to be connected with the endocrine glands. Those who have hyperhidrosis produce excessive sweat, though not necessarily foul smelling. This condition is thought to be glandular also.

Doctors today still use their noses to sniff out the cause of coma in patients brought into hospital emergency rooms. If the patient's breath smells of alcohol, or a poison such as cyanide, the answer is clear. If there is a peculiarly sweet smell, like acetone, the cause of coma is probably diabetes. Should the victim's breath smell of ammonia, the kidneys are most likely at fault; and if a bowel obstruction is present, the breath may smell like excrement. If the patient smells just plain dirty, he or she may be mentally ill, senile, or so impoverished that malnutrition may be the cause of the coma.

Lung and skin abscesses give off a fetid odor and physicians and nonphysicians alike have reported that some patients, as they approach death, smell of pine.

Because we rely so much on fancy devices in our world, young physicians in training are not taught olfactory detection. As a result, some lifesaving diagnoses may be missed.

Dr. Thomas E. Cone, Jr., of the Harvard Medical School, writing in the medical journal *Pediatrics* in 1975, urged his fellow physicians to use their noses as an essential part of the physical examination of patients. He emphasized that statements by mothers about the peculiar odors of their infants should be taken seriously, because a number of inborn errors of metabolism reveal themselves first by odor. In some cases, quick recognition of the problem can prevent damage and death. Three of the diseases cited by Dr. Cone are:

Phenylketonuria (PKU). Described more than thirty years ago, phenylketonuria is manifested by an inborn error of metabolism which prevents the child from manufacturing the enzyme which "digests" phenylalanine. Consequently, the chemical builds up in the baby's body, and the child becomes brain damaged unless measures are taken to reduce amino acids in the diet. The buildup of phenylacetic acid in the urine and sweat gives the child a "mousy" odor.

Isovaleric acidemia syndrome. The children afflicted with isovaleric acidemia syndrome have an enzymatic block in leucine metabolism which causes a buildup of isovaleric acid in their bodies. The genetic defect makes the infants smell "cheesy," or like "sweaty feet." If they do not receive a special protein diet, they develop episodic acidosis and mental retardation.

Maple-syrup urine disease. In this disease, which was first described in 1954, the baby's urine smells like maple syrup on about the fifth day of life. It is caused by an inherited degenerative brain disease. The baby feeds poorly, vomits, and has seizures. If the infant lives long enough, it will be severely mentally retarded. The disorder results from the buildup of certain amino acids, and, again, reduction of these amino acids in the diet can prevent a great deal of damage.

Among the serious diseases of adults that are signaled by the way they smell, acromegaly, a condition involving a tumor of the pituitary gland, causes a very offensive body odor. The pituitary has a great deal to do with the regulation of growth. The gland secretes a growth hormone, and when it is overactive the facial features become coarse, the hands and nose enlarge, the heart and thyroid are affected, and joint pain may occur. When the tumor is destroyed, the person's body and scent usually return to normal.

Disraeli, the great British prime minister, was known not only for his parliamentary skills but for his bad body odor. Doctors who have studied the descriptions given by the medical men of the time believe Disraeli suffered from ozena, a chronic disease accompanied by a fetid discharge from the nose. The inside nasal passages dry out, forming crusts, and there is progressive atrophy of the tissues. Its cause is unknown. It is more prevalent in some families and ethnic groups and is more frequently found in women than in men by a margin of three to one. It occurs in persons with abnormalities in the shape of the skull, malformation of the nasal passages, and/or of the palate of the mouth. It is sometimes associated with atrophic vaginitis in girls, again illustrating the correlation between olfactory organs and the genitals. Ozena may also be accompanied by headache, mental depression, and apathy, all of which could have either a physiological or an emotional cause, or both.

Disraeli's ailment is relatively rare, but all of us, at one time or another, suffer from bad breath. According to the commercials, if we use a certain mouthwash, dentifrice, breath mint, or spray, we'll have "clean breath." But there is more to it than that, and no perfume can long cover bad breath. In order to sweeten your exhalations you must get to the root of the problem.

Fetor ex ore is the medical term for really foul-smelling breath, which is generally caused by a bowel obstruction or some other digestive malfunction. Fetor hepaticus, which results in a similar foul smell, is caused by liver disease leading to the buildup of aromatics in the blood.

Most often, however, bad breath is the result of ingested foods, decaying teeth, unhealthy gums, or respiratory-tract problems. Long-standing mouth breathing, whether caused by nasal blockage or merely a bad habit, can create offensive breath by drying out the normal secretions and facilitating the growth of microorganisms. Infections in the nasal cavity can be at the root of bad breath. Chronic low-grade inflammations of the nose and upper throat may lead to the destruction of the normal "tissue hairs," the cilia, that act as a conveyor belt to remove mucus, odorants, and bacteria.

If you can resist garlic, alcohol, or other tattletale ingestibles you can often avoid offensive breath. If you can't resist them, experts recommend brushing the teeth and tongue and then rinsing out the mouth with a fragrant mouthwash to remove odiferous particles which might still cling to tissues and teeth. Getting rid of alcohol breath is somewhat more difficult. Again the mouth must be cleaned of odiferous remnants. Then nonalcoholic, noncarbonated beverages, such as tea or water, should be drunk to help clear the mouth and throat of volatile odor-bearing oils. Chewing gum or hard candy should be kept in the mouth to encourage the free flow of saliva.

If after taking these precautions you still notice people constantly backing away from you when you talk, better check with your physician and dentist to determine the cause of the problem.

An equally embarrassing antisocial condition is caused by the release of foul-smelling air from the rectum known as flatus. Intestinal gas has been mentioned in medical treatises since Hippocrates described the "flatuosities." Two modern physicians who have been studying the problem are Dr. James L. A. Roth, professor of clinical medicine at the University of Pennsylvania, and Dr. Michael D. Levitt, associate professor of medicine at the University of Minnesota. They became interested in the subject because in their experience, after the common cold, excessive intestinal gas is the most frequent complaint of patients.

Doctors Roth and Levitt said in an interview which appeared in *Medical World News* in April, 1975, that the most important features of diagnosis are recognizing gas-induced symptom patterns and excluding organic disease. They estimated that about 30 to 50 percent of intestinal gas is due to fermentation and the consequent gaseous products of bacteria acting upon undigested food residues. The remaining 50 to 70 percent of gas is from swallowed air.

For all practical purposes, the professors asserted, there are only two gases you can swallow—oxygen and nitrogen. If there are other gases in your gut, they came from fermentation. There may be specific intolerances which exacerbate the problem such as an allergy to cheese or anything with milk or sugar in it. Excess gas may also be produced in the bowel because of incomplete digestion caused by pancreatic insufficiency or wheat intolerance.

Some of us react to emotional stress and turmoil by swallowing air. Giving up chewing gum, candy mints, and carbonated soft drinks can help us get rid of a lot of gas.

The two scientists reported that none of the five major gut gases, oxygen, nitrogen, hydrogen, carbon dioxide, or methane, has any odor at all. Quantitatively, the odiferous components of flatus are unimportant—much less than 1 percent. They include ammonia, hydrogen sulfide, volatile amino acids, short-chain fatty acids, and very malodorous amines such as indole and skatole— which, ironically, are used in the manufacture of commercial perfumes. (Diluted indole is used as a violet-type scent, and skatole is used as a fixative.)

Dr. Roth and Dr. Levitt have found that the problem in studying these smelly trace constituents is that the nose is over a hundred times more sensitive than the best detection machinery in the laboratory.

What about the folklore on the bad effects of eating beans? It's true. Researchers have shown that people on a seven-day diet with beans providing 57 percent of the calories ingested had a marked increase in intestinal gas. (The same holds true for broccoli, cabbage, cauliflower, brussels sprouts, turnips, and sometimes cucumbers, radishes, and onions.) The late Dr. F. R. Steggerda of the University of Illinois discovered that some people don't have the enzymes to digest a certain sugar in beans so that the sugar goes into the large intestine where it is fermented and produces a lot of gas.

Hydrogen, which blows up our gut as it does a balloon, doesn't occur in germfree laboratory animals. But if these animals are exposed to bacteria, they produce the gas within twelve hours. The same phenomenon occurs in newborn babies. At first, since infants have no bacteria, they don't emit hydrogen, but within twelve hours they do. So apparently the only source of hydrogen production in our gut is bacteria.

One of the worst offenders in producing gas, it has been discovered, is apple juice; ironically the very drink most often given to postoperative patients, who usually suffer from painful intestinal gas without this dietary help.

The methane in flatus is produced by bacteria, but it doesn't seem to have anything to do with diet. According to Dr. Roth and Dr. Levitt, about two thirds of the adults in the United States produce no methane, while the other third produces relatively large amounts, up to 0.5 ml per minute.

Methane production seems to be a family trait. The two researchers studied fifteen sets of identical twins and found only one pair that didn't match in methane production. They also studied children in a state hospital for the mentally retarded and found that virtually all the youngsters excreted large quantities of methane. Since they had no genes in common, the researchers reasoned methane production must be transmitted environmentally in the general population. The vector appears to be the mother's intestinal bacteria. Children tend to take on the methane-producing bacteria of their maternal parent, so if your husband has a gas problem, blame your mother-in-law.

Dr. Roth and Dr. Levitt conclude that the odiferous gases we expel are not swallowed but are probably produced by bacteria acting on fats in our digestive tracts. As for what to do about pathologically foul flatus, they suggest antibiotics or *Lactobacillus acidophilus* (which is in yogurt) to change the bacteria in the intestinal tract.

The professors said there is no way to sanitize or deodorize the normal flatus. We're all gassy, but we do not all complain because if the resulting fart isn't noisy or foul we don't feel embarrassed. Unfortunately, if you are embarrassed about passing gas, you become tense, swallow more air, and perpetuate your problem.

Flatus is not the only antisocial body odor to emanate from the genitoanal region. A leaky bladder or a weak anal sphincter muscle may be embarrassingly obvious and, of course, requires prompt medical attention. Less obvious but still needing medical help are the various vaginal discharges which cause malodors.

Almost every woman at one time or another suffers from a vaginal itch, or discharge, or both. The condition may soon disappear, or it may become increasingly worse. Vaginal infections are common and occur in about one third of the women of childbearing age. It has been reported that 95 percent of the cases today can be attributed to *Trichomonas vaginalis* (a protozoan infection), *Hemophilus vaginalis* (a bacterium), or *Candida albicans* (a yeastlike fungus infection, more commonly referred to as monilia).

The most common of these disorders today is monilia. It is believed to have increased in incidence because of oral contraceptive pills, which contain hormones that change the pH—the degree of acidity—of vaginal secretions. The less acid these are, the more favorable the climate for the growth of causative organisms. Antibiotics are also blamed for the increase in vaginal infections. They kill staphylococcus and other disease-producing bacteria, but they also destroy benign bacteria which protect the vagina against harmful germs.

Vulvar odor is also sometimes associated with the wearing of close-fitting underwear or pantyhose, particularly those made of nylon. Physicians often recommend that women wear cotton instead of nylon because the cotton is more absorbent and allows a freer flow of air. Pantyhose manufacturers have been made aware of the problem, and a number of them now produce nylon hose with cotton crotches. Pantyhose and nylon panties cannot by themselves cause vaginal infections, but they do provide a warm, moist environment in which the causative agents can multiply.

Bubble bath has also been reported to be a frequent cause of lower-urinary-tract inflammation and, occasionally, of vaginal discharge. So have perfumed soaps, talcs, bath oils, bath salts, and feminine hygiene sprays. The intense itching and foul-smelling discharge of a vaginal infection can have dramatic psychological effects on women, according to the medical literature. Irritability, nervousness, and a wide variety of other apparently unrelated symptoms have been reported.

Genital infections can be passed back and forth between sex partners and may cause misunderstandings between them, sometimes about the possible source of infection.

A physician can easily diagnose the exact cause of vulvar infection and odor by taking a smear. The best course of therapy can be prescribed, and the infection with its telltale malodor will soon disappear.

The foot is another part of the body which is often odoriferous. Many animals have special scent glands in their paws, so they can mark their territories as they walk. Our own feet are so odoriferous that our scent penetrates leather shoes, enabling a dog to follow it by sniffing the ground where we walked as long as two weeks before. Even if your nose is not as acute as a bloodhound's, you have only to open a closet door where shoes are stored and you can immediately detect the scent of their owner's feet.

Our feet carry the entire weight of our body when we walk. With each step we take, we jolt them with a force amounting to several tons. The dark, moist environment of our shoes provides an ideal place for bacteria to work on the sweat from the glands of our feet. Some people's feet produce more sweat than others' and environmental and emotional factors can also affect the amount of perspiration.

The odor emanating from your feet can usually be combated by simple hygienic methods. If your feet perspire excessively, bathe them frequently with alcohol or witch hazel. If that doesn't work, a commercial antiperspirant can be purchased in a supermarket, drugstore, or shoe store. Such products usually contain aluminum salts, which temporarily reduce the amount of sweat on the skin.

There are oral medications prescribed by physicians which reduce the amount of sweat you produce. Foot specialists have devices in their offices that dry the sweat glands in the feet, but this should be done only as a last resort.

Just as the feet, palms, and underarms sweat excessively when we become emotionally aroused, so too does the whole body. People who are extremely anxious and frightened often sweat profusely and can even lose control of their bladders and bowels. Some may show no outward signs but give off a peculiar scent which has been said to be instantly recognizable by animals. There are numerous reports that the odor generated by fear in the human body stimulates a dog to attack and makes a horse unmanageable.

Misery supposedly makes us smell bad, and happiness gives us a sweet odor. According to Havelock Ellis, in churches where the religious excitement is high there is a pleasant perfume—an odor of sanctity.

Changes in scent in response to fear are obvious among animals. A resting rattlesnake is said to smell like a newly cut green watermelon, while an angry snake smells like a wet dog. Rats can smell fear in other rats. Experimenters have shown that when a group of laboratory rats is subjected to electric shocks and their only contact with another group of rats is by scent, the second group displays fear and excitement without any other possible cause. Similar experiments have been performed with fish. With no contact but the water exchanged between their two tanks, the shocked fish in one tank signal chemically to the untouched fish in the second tank, causing them to become alarmed.

Perhaps the most intriguing and yet elusive aspect of emotional odor detection concerns the mental disorders called schizophrenias. In victims of schizophrenia the senses often become distorted and the patients hear, see, and smell things in a supersensitive way. A number of investigators have reported that some schizophrenics can apparently smell the moods of other people.

On the other hand, schizophrenics themselves have a peculiar smell. It is accepted among mental hospital personnel that the schizophrenic wards have a peculiar odor which does not come solely from excrement and disinfectant. The unusual scent seems particularly intense in rooms where insulin-shock therapy is being given and appears to come from the skin of the patients.

This distinctive smell is particularly strong in catatonic schizophrenics, who become frozen in positions and seem completely out of contact with the environment. They have unusually greasy skins, and no matter how much they are bathed they smell like skunks.

Dr. Kathleen Smith of the Malcolm Bliss Health Center in St. Louis, Missouri, has been in the forefront of those trying to identify the cause of the schizophrenic's peculiar odor. She reports that the odor becomes stronger when the patients are very ill and less strong as they improve. The odor, she says, seems to be of metabolic origin, but so far its source has not been identified. It has been suggested, however, that it might arise from apocrine sweat or sebaceous gland secretion, and is probably, along with some other symptoms of schizophrenia, caused by some biochemical abnormality.

What makes the olfactory hallucinations of schizophrenics so intriguing is that if they smell things that aren't there, the prognosis for their recovery is poorer than if they have visual or auditory hallucinations alone. Furthermore, such smell hallucinations are more common in schizophrenics with delusions of sexual change.

One schizophrenic will, after a meal, hallucinate the sexual smell of the cook who prepared the dinner, while another will see a jet plane and hallucinate holy, heavenly scents. But the odor hallucinations of schizophrenics generally are unpleasant, and they often jam things up their noses in a futile attempt to stop the odor. Sometimes an unpleasant odor hallucination will cause a physical reaction, such as grimacing, nausea, vomiting, or even fainting.

For some reason the incidence of nasal disease or malformation in patients with smell hallucinations is greater than in those without them. The nasal pathology—when it is not connected with actual injury from objects being inserted into the nostrils—seems physiologically unrelated to the incidence of hallucinations. However, it is theorized that, as a consequence of altered thinking, the pathological changes in the nose act as a trigger for the precipitation of hallucinations.

Diagnosis of mental ills, such as the schizophrenias, is an educated guess. So too are other diagnoses. If physicians could develop instruments which can detect

Sickly Smells 71

nature's minute chemical changes in such illnesses, earlier and better diagnoses could be made.

Among the pioneers in the field are Dr. Andrew Dravnieks and Dr. Boguslaw Krotoszynski of the Illinois Institute of Technology's Odor Science Center. They are working on various devices to gather odors for analysis. With all of them, air introduction and withdrawal must be carefully controlled. The body vapors are collected and then analyzed with the gas spectrograph as well as by expert human sniffers.

They use a special mouthpiece to collect vapor samples from the mouth and lungs; samples of vapors from the skin are taken using a Teflon cup. Vaginal vapors are sampled with a perforated Teflon insert, and whole body vapors are gathered by placing a person in a device like a giant glass test tube on a Teflon-lined stretcher.

The Chicago researchers seal a person in the glass-tube device for forty-five minutes. The air is then analyzed. On the basis of these odors the machine can report what sort of food the subject has been eating, what sort of fumes he has been breathing, and even his race. The machine can identify individuals with 80 to 90 percent accuracy from their "odor signatures."

Conceivably, police could adapt Dr. Dravnieks' and Dr. Krotoszynski's device to sniff the air at the scene of the crime to determine the sex, race, eating habits, and even the occupation of the criminal or criminals.

In another project, designed to determine the malodorants produced by the human vagina, volunteers were enlisted to agree to follow a prescribed routine. They had to refrain from douching for seven days and were required to avoid intercourse for forty-eight hours prior to testing. For twenty-four hours before testing they could not bathe or shower, nor could they eat heavily spiced foods containing garlic. They could not use perfume, perineal powder, or vaginal hygiene sprays. A total of six vaginal odor samples were obtained on two days, during various times of the day. Both humans and mechanical devices were used to detect odors. It was determined that there were from twenty-two to eighty-seven odorous substances but only about fourteen were malodorous. Seven of those malodorants appeared more often than the others.

The Chicago researchers could not identify the source of the specific malodorants but theorized that they were probably decomposition products of microorganisms, desquamated cells from the urogenital tract, normal vaginal secretions, and sperm.

Some British scientists have suggested using dogs to make diagnoses in medicine. The great olfactory acuity and scent memory of dogs, they reasoned, could be used to train the animals to detect abnormal metabolics in sweat, blood, and urine. Dogs might learn to detect schizophrenias and other biochemical maladies. If dogs proved too expensive, then the British physicians suggested eels, whose olfactory organs are also supersensitive to chemical changes.

Not to be outdone, American scientists have already developed mechanical "people sniffers," which were used to pinpoint enemy troops in the jungles of Vietnam. The Israelis improved upon the American version, and their "sniffer" can be aimed at the ground from helicopters and smell the enemy by detecting traces of chemical salts emitted by perspiration, urine, or feces. Such devices are thought to be adaptable to ferreting out hidden caches of heroin and other illegal drugs, as well as money and minerals. With the aid of computer attachments, it is conceivable that they could one day be used to rapidly and accurately diagnose human diseases.

5 When the Nose Doesn't Know

JUST AS a person's body odor can lead to the diagnosis of ailments, so too can an individual's own ability to smell. Olfactory malfunctions signal trouble in the body because smell is basically a chemosensory system and if there are major changes in body chemistry, the nose often knows first.

An estimated two million Americans have lost their sense of smell or taste, or both. The absence of these two senses, while not as devastating as the loss of sight or hearing, can take much of the pleasure out of life. Furthermore, one fourth of the patients who lose their sense of smell also lose their desire for sexual relations.

Loss of smell can have fatal consequences. Newspapers carried the story of a civil engineer who had a decreased sense of smell. He tried to unplug a sewer line by opening a manhole cover and pouring in the appropriate chemical. After waiting more than half an hour he opened the manhole to see if the chemical had done its work. Unable to smell the toxic fumes which had accumulated, he went down into the manhole to take a closer look and was soon asphyxiated.

Truck drivers have lost their lives because they could not smell poisonous gases leaking into their cabs, chemists have been asphyxiated in their laboratories, and jet mechanics have died because they could not detect dangerous vapors.

Smell and its sensory companion, taste, affect the functioning of the digestive system. Stimulation of these senses can initiate eating, influence the volume and character of saliva flow, increase intestinal motility, modify both the volume of pancreatic flow and its protein content, and influence the selection of nutrients.

Dr. Robert I. Henkin, who formerly headed the section on neuroendocrinology at the National Heart and Lung Institute in Bethesda, Maryland, and now serves as director of the Center for Molecular Nutrition and Sensory Disorders at Georgetown University in Washington, D.C., is a specialist in the treatment of smell and taste disorders. In fact, he is one of the few clinicians in the world skilled in treating patients for such problems.

Dr. Henkin first became interested in the field when he noticed, while working as a medical student at Columbia University's College of Physicians and Surgeons in Manhattan, that many hepatitis victims lost their senses of taste and smell. He wondered why this should happen to patients suffering from a viral disease affecting the liver. He discovered that as soon as hepatitis victims began to get better, their senses of taste and smell did too; in fact, improvement of olfactory perception was a simple guide to the course of the patient's illness, an indicator of systemic recovery.

A pioneer in this field, Dr. Henkin discovered that there were decreased concentrations of zinc in the blood of hepatitis victims, an observation he began to pursue in victims of taste and smell disorders from other causes. A zinc deficiency is often found in pregnancy and in postoperative and burn patients. Dr. Henkin wondered whether the loss of zinc during surgery could be responsible for the lack of appetite in postsurgical patients. Is zinc necessary for the maintenance of olfaction and taste?

An important trace element found in the human body, zinc is normally obtained in the diet in meat, seafood, eggs, legumes, and whole-organ products. A deficiency of this element has been implicated in a number of disorders, including dwarfism and certain types of infertility.

The pieces of the puzzle are beginning to fit together. Dr. Henkin and his colleagues say that smell acuity is a particularly important diagnostic tool for the functioning of the sex glands. Tests of hormonal levels in teenaged girls who have failed to menstruate may be inconclusive simply because they have not yet reached puberty. But if their olfactory acuity is normal, the girls will most likely have ova and will eventually menstruate; if their sense of smell is poor, they probably have no ova and should be placed on estrogen therapy at the appropriate time, during puberty.

The same finding holds true for young boys. If their sex glands are underdeveloped and they have no sense of smell, they will probably be sterile and should be placed on testosterone therapy at puberty. If a boy has some sense of smell, rather than none at all, he is probably normal and will eventually be fertile. According to Dr. Henkin, the smell acuity test is about 80 percent accurate for both sexes.

The nerve tracts serving olfaction and the functioning of the ovaries and testes, collectively known as the gonadal system, run in the same area of the brain. In fact, researchers have been able to demonstrate that in animals the development of the olfactory bulbs and the development of the sex-gland system are related basically to a hormonal effect on the gonadal system. Zinc has also been reported to be necessary for normal gonadal development.

Since zinc has been found to be low in many patients with an inability to smell, it should be easy to treat them with zinc, and Dr. Henkin often does give large doses of the element—about 100 mg per day—which are effective in about one third of the patients, seeming to benefit only those with a specific zinc deficiency. There is no definite understanding of why zinc is important to smell, but Dr. Henkin theorizes that it is probably necessary in the fluids that bathe the olfactory receptors, as well as in saliva. His findings are that some patients get better with zinc therapy, some with placebo pills containing no drugs whatsoever, and some without any treatment at all.

The Georgetown physician observed that some people are born with a lack of smell or taste or both, but that in most cases people with these dysfunctions were formerly normal. "People who have lost their sense of taste or smell and realize that it may be that way for the rest of their lives feel depressed," he said. "They've lost something of great value and are very unhappy. It is not unlike the loss of a limb."

Dr. Henkin reported that about 10 percent of the patients who come to his clinic are found to have undiagnosed malignancies. Rapidly growing tumors, he theorizes, may rob the body of zinc and therefore affect olfaction. Taste and smell dysfunctions, together with frequent and severe headaches, may signal a brain tumor or a malignancy elsewhere within the system.

Loss of smell may be associated with nasal polyps, severe burns, head and neck surgery, dentures, and heavy smoking. It can also be caused by a stroke, muscular disorders, or infectious diseases such as diphtheria and encephalitis.

A large percentage of patients who lose their sense of smell have had head injuries. Many patients who suffer blows to the head—even minor ones—lose their sense of smell and taste, or these senses may be impaired. The problem is that the olfactory symptoms may not appear until several weeks or months after the injury occurred. In loss of sensation caused by head injury, olfactory acuity returns spontaneously in almost a quarter of the cases.

A whopping 40 percent of the patients Dr. Henkin treats at the Georgetown clinic have suffered from an influenzalike illness. Such patients report a severe flu followed by lingering fatigue and a diminished sense of taste and smell. After a few weeks, the patients realize they are feeling better in general but they are still unable to taste or smell normally.

Some of the patients eventually get better spontaneously, some respond to zinc therapy, and some never regain their lost senses of taste and smell. Some female patients who suffer from this postinfluenza loss of acuity also have a concurrent onset of menopause and in men the loss of smell may be accompanied by decreased beard growth —again suggesting a link between the sexual and olfactory systems.

The importance of disturbances of taste and smell in clinical medicine is just beginning to be appreciated, in a large part thanks to Dr. Henkin. Patients who come to the Georgetown clinic—and there is a long, long waiting list—are asked to give complete medical histories and to submit to a series of complicated tests. Body fluids are tested radioactively and other electronic equipment is used for fluid and vapor analysis. There are not only numerous causes of smell loss but also various degrees and manifestations of the disorder:

Anosmia is the technical term for the complete inability to detect or recognize any smell.

Dysosmia means any distortion of normal odor perception.

Cacosmia is the term for a condition in which an obnoxious smell is perceived on inhalation of normally pleasant-smelling odorants such as perfumes, soaps, hair sprays, and food.

Phantosmia designates the hallucinating of a variety of odors, pleasant or unpleasant, which are smelled intermittently or persistently by the patient though no apparent odorants are present. Schizophrenics have such odor hallucinations, and so do victims of temporal-lobe brain tumors.

Heterosmia means the perception of an inappropriate smell of a consistent nature. This smell is unusual and unexpected but not necessarily foul or obnoxious.

People find such aberrations very disturbing. Even if patients smell roses all the time, they are bothered that they are smelling things that others are not. Sometimes they believe that they have a more acute sense of smell than other people. These symptoms may be the result of a variety of conditions, including liver diseases, metabolic dysfunctions, and brain tumors.

Certain people with smell disorders exhibit an "extinction" symptom. When they begin to eat the first whiff or taste of the food is all right, but after that their sensory ability disappears. This sudden loss is extremely frustrating to them.

Dr. Henkin has identified two more categories of smell disorders: Type I hyposmia and Type II hyposmia, based on his observations of patients whose sense of smell has been affected by surgical procedures, involving the oral, aural, and nasal regions as well as the brain. He studied two groups of patients before and after two specific surgical procedures—one for cancer of the sinuses and the other for removal of the larynx.

In seventeen patients operated upon for cancer of the sinuses, the lining of the nasal cavities with its nerve cells was cut and so were the olfactory nerves leading to the brain. The paranasal sinuses were reamed out. For a short time after surgery, Dr. Henkin and his group discovered, all the patients lost their ability to detect or recognize any smells whatsoever. As time passed, they gradually recovered their sense of smell, although its acuity was greatly diminished compared to normal subjects' and to their own presurgical ability. Before the operation, the odors had been sensed and recognized in the upper portion of the nose, the primary olfactory area; after surgery the patients felt vapor sensations in the nerves of the cheek and throat. This surprised the researchers, since they had assumed, as had everyone else, that olfaction occurs solely in the nose.

It has long been known that in lower species, such as the reptiles and herbivores, the throat and mouth have an olfactory sense so that animals can smell while feeding. In fact, Dr. Henkin points out that in the Tarzan films of the 1930s the crocodiles coming toward Tarzan in an African river with their mouths open were not only trying to eat Tarzan but to find him. A crocodile's eyes are set far back in its head, and its visual sense is not as developed as in many other creatures. Its olfactory system is keen, however, and therefore the crocodile relies heavily on the sense of smell for finding food. The epiglottis evolved in such animals to prevent food from going into the lungs. In man, this separation is less complete, but the epiglottic tip may still serve an olfactory function.

Hence, Dr. Henkin has categorized as Type I hyposmia the total absence of responsiveness in the primary area of olfaction—the nose—with the preservation of responsiveness in the accessory areas of smell in the mouth and throat. Type II hyposmia designates only decreased acuity at the primary area of olfaction with maintenance of acuity in the accessory areas of olfaction.

A type of operation which one would think should not interfere with the sense of smell but does, is the removal of the vocal cords—laryngectomy. Immediately after surgery and for as long as eight years afterward, patients lose their sense of smell. Even after their ability to smell is recovered, their acuity does not return to normal. If such patients are not warned about the loss of smell, as well as the loss of the ability to speak, it can be a terrible shock, Dr. Henkin reported. The Type I hyposmia observed in laryngectomy patients is similar to that in the patients who have sinus surgery with their olfactory nerves cut. They can no longer smell with their nose but appear to retain some sensation in their mouth and throat. They can be taught to increase their olfactory acuity by a certain way of inhaling.

Why there is a loss of smell after laryngectomy is not clear. The loss of smell apparently is a complication of surgical interference with the voice box rather than a result of any significant alteration in airflow to the olfactory receptors.

One theory is that interference with the sensory nerves of the larynx during surgery may alter the sense of smell through a complex feedback mechanism or by interference with some diffuse anatomic protection system from the larynx to the brain. Smell and respiration are intimately tied together, and it is known that there are certain reflexes present in the nasal cavity. Patients with sex-hormone deficiencies along with olfactory disorders also have impaired pulmonary function. No one is sure why these diverse systems are so interrelated, but research to find the answer is in progress.

It is easier to understand how the lack of a sense

of smell affects food intake. Dr. Henkin points out that when all foods tend to taste alike and the only distinction is in texture, eating can be boring. More serious, however, is when the disorder makes food smell bad. With this type of cacosmia, patients tend to avoid proteins, which smell particularly foul to them but which are vital to health. Alcoholics frequently suffer from cacosmia. They have a vital need for protein but protein-rich food smells terrible to them. These victims may end up eating the same food for every meal and become malnourished. When cacosmia occurs in children, parents often take them to physicians who in turn recommend behavioral therapists, instead of experts who could help with the very real physical condition.

Another diet-related condition has recently been coupled with loss of taste and smell—the lack of sufficient thyroid hormone, a hormone involved in the metabolism of food. In most cases, patients suffering from diminishing thyroid-hormone secretion are not aware of the gradual loss of their ability to smell and taste. They sometimes begin to add so much salt and sugar to food that they may adversely affect their own health and, if they prepare the meals, that of their family as well.

When the thyroid gland slows down, the patient—usually a woman—may feel more tired than usual, losing her pep and displaying little interest in her daily activities. She is lethargic. Her hair seems dry and coarse, and hair loss occurs. Her complexion looks pale and sallow. She may feel cold all the time. Her menstrual periods may be irregular, and she may be constipated.

Patients with underfunctioning thyroid glands show very little interest in eating and frequently complain of loss of appetite. It is a common misconception on the part of the general public, according to Dr. Richard S.

Rivlin of Columbia University, who, with Dr. Henkin, made the correlation between smell and thyroid deficiency, that an underactive thyroid gland or slow metabolism is the cause of obesity. In overweight patients, Dr. Rivlin said, the thyroid is nearly always normal. His present research may, in fact, provide an additional explanation of why hypothyroid patients don't eat—food just doesn't taste very good to them.

After treatment with thyroid, he said, every patient reported appetite improvement. In most cases, the patients noted themselves that their smell and taste sensations markedly increased. This improvement was confirmed by objective testing and showed that in most instances the patients' olfactory and taste abilities returned to normal. One seventy-three-year-old woman fully recovered her senses of smell and taste after only sixteen days of therapy with thyroid hormones.

With arthritis, again particularly with women victims, there is a tendency to develop Sjögren's disease, a condition causing abnormal glandular secretion, particularly those glands lubricating the eyes, nose, and throat. Patients with this syndrome feel miserable, have dry noses, and have a decreased ability to smell.

Another serious disease, Wegener's granulomatosis, also causes dryness and crusting of the nasal passages with decreased olfactory acuity. Anyone noting unusual dryness of the nose and throat over a period of time should seek medical advice.

In still another glandular disorder, the opposite effect occurs. When the adrenal glands, which lie just over the kidneys, begin to malfunction, the sense of smell is so sensitized that it becomes not only annoying but diagnostically significant. People suffering from Addison's disease, the malfunction that affected the late President

Kennedy, are able to detect odors about 100,000 times more acutely than healthy controls can. When Addison's disease victims are given adrenal hormones, their sense of smell returns to normal.

Why the lack of adrenal hormones causes hyperosmia can be explained by the role these hormones normally play in the control of nerve activity. When such hormones are deficient, release of the inhibition they impose would occur, making the nerves supersensitive.

Nothing can equal the harmonious workings of our senses in giving us a feeling of well-being. We do not really appreciate them until we begin to lose them. Unfortunately, little medical attention is paid to our pleasure senses, smell and taste. There are routine tests for 20-20 vision and for hearing, but tests for olfaction are rarely, if ever, performed.

Dr. Henkin has developed two test kits, one for smell and one for taste, which are awaiting Food and Drug Administration approval. If he receives it, he will distribute these kits to other physicians, who can systematically examine the olfactory and taste ability of their patients as a part of diagnosis. He hopes to train many other physicians in the art and science of treating smell and taste disorders. He works now from 7 A.M. to 7 P.M. every day and feels overwhelmed by the numbers seeking his help.

It is well known that many people are made ill by certain smells. Albert Weber, the supersmeller who detects rotten fish for the U.S. Food and Drug Administration, is allergic to the odor of rotten frog's legs. Before he examines a potentially bad batch, he is forced to take an antihistamine tablet.

Many people are allergic to the odors of paint, strong perfumes, and smoke. Victims of vascular-type headaches are often warned by their physicians to avoid such smells to prevent an attack.

Vile smells, such as stale vomitus, can make almost anyone sick, but smells can also be therapeutic. In fact, the psychiatrist Dr. Ralph Crawshaw, writing in *Prism*, noted that it is often the smell of the medication rather than the medicine itself which helps to make a patient better. He firmly believes that certain smells can make you well.

There are a lot of people out there who need help, not only for professional reasons—some are cooks or florists who need their senses for work—but because much pleasure goes out of life for those who cannot smell or taste anything.

Part II:
The Success of Sweet Smells

6 Perfume Politics

THERE IS emotional magic in perfume.

Before a scientist ever lifted a test tube or a patient lay on a psychiatrist's couch, human beings knew that moods could be changed, memories brought back, evil masked, sexual desire aroused, and life made generally more pleasant by the use of sweet-smelling scents.

Fragrance, it has been said, is the openly broadcast gift of plant life. It occurs not only in flowers, leaves, and fruits, but in all the creatures of the earth, though we may not recognize it as such. Fragrances, in the marvelous design of the world, are the lures which attract procreators and induce them to perform the function which will assure future generations. Whether it's the bumblebee spreading the pollen or the tomcat howling after his ladylove, a specific scent is irresistible to them.

Nature is the greatest chemist of all. She creates complicated mixtures of special design. But we humans, superior creatures that we are, have done a superb job of using her materials for our own purposes.

The word "perfume" literally means "through smoke." The history of the use of scents begins with primitive man's discovery that burning certain woods and resins made things smell better. In early temples of worship, when human and animal sacrifices were made, burning aromatics covered the slaughterhouse smell of blood and flesh.

More than 5,000 years ago, ancient Egyptians burned sweet-smelling fragrances to the sun god, Ra, as he made his daily journey across the sky: resins, the exudates of plants, were used as he rose in the east; myrrh, the sap from an Arabian tree, was used when he was directly overhead; and a mixture of sixteen ingredients was sent heavenward as he set in the evening.

The Egyptians also used perfumes for anointing the favored and for embalming the dead. Dried perfumes have been found in the pyramids. One, nard, is believed to have been valerian, extracted from a root and still used today in perfume manufacturing. Another, lotus, is thought to have been extracted from a sweetly perfumed water lily. The Egyptians created unique scents for each mummy so that should the dead person be separated from any of his or her parts such pieces could be recognized immediately in another world as belonging to the whole.

Egyptians taught the art of perfumery to their slaves, the Hebrews, who then recorded the use of many aromatic materials in their sacred books. Among the scents described were jasmine, from a white flowering shrub, rose, from the flower, and animal products such as musk.

When the Hebrews left Egypt, they took with them the knowledge of the power of perfumes and the formulas for making certain mixtures. They began to trade in aromatics. In 1700 B.C., according to the Bible, the Ishmaelites came from Gilead with their camels bearing spicery, balm, and myrrh. Spicery is believed to be either storax, the resin from the bark of an Asiatic tree, or tragacanth, the gum of an Asiatic plant. Both are still used in the perfume industry today.

Perfume Politics 89

According to the book of Exodus, the Lord told Moses, who had departed from the land of Egypt, to take five hundred shekels of pure myrrh, two hundred fifty shekels of sweet cinnamon, two hundred fifty shekels of sweet calamus, five hundred shekels of cassia, and some olive oil. The ointment made from these ingredients was declared holy and was used to anoint the tabernacles of the congregation.

For thousands of years, priests were also doctors, and as they used new substances for temple offerings, they discovered many aromatic medications. It is thought that the intoxicating properties of burning hemp were first found in this way.

Perfumes were very costly and held in high esteem. Perfume wars were fought both on the battlefield and in the bedroom. It is, therefore, not surprising that they were used for purposes other than religious and medicinal. Joseph's brothers received perfume and spices as payment when they sold him; and when the beautiful widow Judith sought to save her people from annihilation by the troops of Nebuchadnezzar's general, Holofernes, she used perfume. Attracted by her fragrant body, Holofernes was lured into his tent, where sweet-smelling Judith cut off his head.

The Queen of Sheba also used perfumes to conquer. When she visited Solomon about 800 B.C., she brought him fragrances and successfully seduced him, another landmark in the power of aromatics.

Babylon, Nineveh, and Carthage became great centers of perfumery in the seventh century B.C. The inhabitants collected aromatics—odorous gums from Arabia, camphor from China, and cinnamon from India—and exported them through Phoenician merchants to the entire world. The use of perfumes in Assyria reached its zenith in 650 B.C., when the somewhat peculiar ruler Ashurbanipal

THE SMELL BOOK

dressed himself up like a woman, used cosmetics lavishly, and soaked himself in perfume.

In 500 B.C. Hindu temples were being built with fragrant sandalwood. India was richly endowed with all kinds of scented flowers, woods, resins, and musk. Their perfume was originally kept, as in the Middle East, strictly for religious rites. Later, humans began to perfume themselves and their homes with rose, jasmine, tuberose, narcissus, and sandalwood. Heavenly scents were common at ceremonies and feasts. An ancient Indian custom, which still persists today, prescribes that the young bride-to-be of an important personage must bathe for several hours in a perfumed bath each day of the month before her wedding.

While the Indians were bathing young brides in perfume, the Chinese were literally marking time with incense. They designed openwork seals of metal which were placed on carefully tamped beds of ashes. Into the holes, they poured incense. When the seals were removed, the incense lay in an intricate pattern with marks designating the hours. As the incense burned, one had but to look and observe its progress to tell the time of day.

The Chinese also used a coil of incense with hourly markings. Later, the Japanese improved upon the device and used different fragrances so that the day would have an appropriate aroma for each of its moods. So accurate were the incense clocks, called joss sticks, that geisha girls would say they worked "six sticks" or however long their day had been.

The Japanese have a legend that burning incense attracts spirits called Jiki Ko Ki, who come to eat the smoke. Such spirits are those of men who sold bad incense during their earthly lives and were punished by having smoke as their only food through eternity.

Perfume Politics 91

The ancient Greeks learned the art of perfumery from the Asian countries. Hippocrates, the most famous physician of all times, outlined the study of the skin and advocated not only healthful living habits but specially scented baths and massages. He also recommended perfumes as medicines for certain diseases.

The Greeks also believed that fragrances could prevent drunkenness and other maladies. They used them on flags, linens, walls, horses, and dogs. In fact, the excessive use of perfume for all social occasions led Solon, the famous Athenian lawgiver of the sixth century B.C., to prohibit the sale of perfume in his cities. Probably he suffered from allergies to the ingredients in the popular scents.

The Romans were greatly influenced by the Greeks, and perfumes became extremely important during the reign of the great Roman emperors. Essential oils extracted from flowers, leaves, and roots were used in abundance in Roman baths. Perfumes were used in public places, in mansions, in palaces, and were often applied to the walls in a form of paste. The Romans added rose perfumes to their wines and even founded a women's senate whose task it was to test the quality of perfumes in use, a sort of modern consumer panel.

All cosmetics required for the elaborate beauty routines of Roman women were made at home by young slaves called *cosmetae* who were supervised by an older woman called the *ornatrix*. This probably paid off, since Roman men were obviously easily seduced by fragrances. Legend has it that, a few decades before the birth of Christ, Cleopatra arranged for her first meeting with Antony to occur in a room thickly carpeted with sweet-smelling rose petals. And you know what happened to Antony.

When the Christ child was born, the Three Wise

Men followed the star and brought the infant two resins from the Arabian trees, frankincense and myrrh. Frankincense is still used in churches and in the homes of Indians and Arabs today.

Nero, the emperor of Rome in A.D. 54, was also susceptible to scents. No wonder his wife, Poppaea, spent hours at her dressing table. She employed hairdressers and perfumers from Cyprus and used the latest scents made from oil of ambergris, a musky substance produced by sperm whales. At night she covered her face with a paste of thick cream and powder to protect her skin. She used crocodile mucus (is that much sillier than today's turtle oil?) to keep her hands soft. After her scented bath, her slaves dried her with swansdown. Her clothes and her jewels were kept in perfumed boxes.

Nero himself was not only a fiddler, but a great admirer of both perfumes and his wife. When she died, he reportedly used more incense at her funeral than all Arabia could produce in ten years.

As the Roman Empire declined and Europe sank into the Dark Ages, the Asian and Middle Eastern cultures flourished and fragrant materials were considered as precious as jewels. Unlike the austere European Christians, Middle Easterners liked sensuous materials. Muhammad in the seventh century A.D. made no attempt to forbid the use of scent. In fact, he said the three things he liked best in this world were women, children, and perfume. His paradise was filled with fragrances.

Mosques, buildings of awesome beauty, were always made even more appealing by the addition of a small quantity of musk to the mortar. It is said that even today, when the sun shines on a mosque, if you sniff the air, you can still detect a hauntingly delicate scent of musk.

Perfume Politics 93

In the ninth century the Danes tried to reintroduce bathing into England, but their efforts did not meet with much success. The Europeans at this period were culturally far behind the Asian and Middle Eastern peoples, and the designation of the Dark and Middle Ages as "a thousand years without a bath" gives you some idea of how people and places smelled. Europeans had no drains, little soap, and a marked distaste for bathing. They lived in a stinking environment, where the odors from the castle drains were used to keep the moths from the lord's clothes hanging in the garderobe, which was also a lavatory. There was no refrigeration, and most cattle had to be slaughtered before the onset of winter and their meat crudely preserved; so off-tastes were added to off-odors, leading to a great demand for costly spices.

In the tenth century A.D., the famous Arabian physician, Avicenna, succeeded in isolating from the rose some of its perfume in the form of oil (attar) and was the first to produce rose water, or attar of roses. This, in essence, was the beginning of the extraction of essential oils by distillation, a practice still used today.

The Crusaders left home in the eleventh, twelfth, and thirteenth centuries telling their wives and priests that they were setting out to recapture the Holy Land from Islam. They soon learned to covet the materials which the Middle Eastern countries used to perfume their food, bodies, and environment, and came back with the sensuous harem perfumes and the spices of the lands they ravaged. Among the most popular scents they brought back was red chypre, a mixture of resins and oils which included damask roses, red sandalwood, aloes, cloves, musk, ambergris, and civet—all still in use today. Red chypre was sprayed through a peacock's beak and became more popular than Avicenna's rose water. The

Crusaders also brought back, from Persia, an attractive, slender atomizer, which was used to spray scent on just about everything in sight.

By the twelfth century A.D., perfumes were extremely popular in Europe. A German Benedictine nun, Hildegard of Bingen, supposedly invented lavender water, by distilling canella, wallflower, galingale, and grains of paradise.

Philippe Auguste, in A.D. 1190, became perhaps the first consumer-conscious king. He officially recognized perfumers by granting them a charter. (This charter was reconfirmed by John the Good in 1357, Henry III in 1582, and again by Louis XIV in 1658.) In effect, the charter set standards. It recognized that the art of perfumery was a difficult one and called for years of training. A four-year apprenticeship was required, followed by three years as a journeyman. It was then possible to qualify as a master perfumer. The candidate had to go before a jury—not unlike our professional boards today—to be certified.

A century later, Henri de Mondeville, a Norman, made the first move to separate cosmetics from the practice of medicine. Trained in Bologna, Paris, and Montpellier, he had established a fine reputation as an anatomist and physician. He wrote a long textbook on surgery which clearly distinguished between medical treatment of pathological conditions of the skin and the application of cosmetics merely for embellishment.

The entire period of the Renaissance was one of great creativity and progress in perfumery and cosmetics, chiefly through the rapid development of all commerce and industry. Around 1370 there appeared in Hungary a perfume supposedly created by a hermit for the Hungarian queen, Elizabeth. It was so powerful, the story goes, that despite the fact that she was in her seventies, the perfume made her so desirable that the King of Poland sought her hand. The recipe for this scent, which was called Hungary water, was the first published for an alcoholic perfume. It was a toilet water based originally on rosemary but including many other plants. It was a bestseller for more than four centuries.

In the meantime, the great explorers were doing their share by bringing back the raw materials. In the thirteenth century Marco Polo, a Venetian, established a trade route to the Orient. The discovery of the Western Hemisphere at the end of the fifteenth century opened the vast region to exploitation by Spain. Around the same time, the circumnavigation of Africa by Vasco da Gama gave the Portuguese the advantage of a new route for the coveted traffic in spices from India, while shortly afterward another Portuguese, Fernando Magellan, sailing from the Moluccas (Spice Islands) under the Spanish flag, discovered the strait at the foot of South America which bears his name. Within fifty years, through the exploitation of their newly discovered territories, Portugal and Spain became the wealthiest nations in the world.

The many new plants or variations of familiar ones brought to Spain from the New World were well described by a Seville physician, Nicolas Monardes. Among them were animé (thought to be elemi), bitumen, balsam of Peru, copal, castoreum, chinaroot, guaiacum, peppers, sarsaparilla, storax, and sugar (until then used only medicinally and very expensive).

All this wealth of new aromatic materials fired the growth of an industry which had long been established in Spain. The Arab perfumers, who were granted permission to remain in the country when their fellow countrymen were expelled, jealously guarded their secret formulas. But little by little, the Spaniards succeeded in penetrating their art and soon became as skilled as the Arabs. The art of perfumery quickly passed completely into Spanish hands. Musk, ambergris, and civet were used to scent tanned hides for long periods of time. Then they were made into gloves and belts, the leathery fragrance of which was called Peau d'Espagne, Spanish leather. It is still popular today.

In Italy, too, the perfumers were at work. The Dominican monks of Ste. Marie Novella in Florence established a laboratory in 1508 for making scents. Their lily water became famous throughout Europe, and it is recorded that rich people could order custom-made scents from them.

Frenchwomen were still buying their perfumes from Roman merchants whose ancestors had come to France with the troops of Julius Caesar, but that was soon to be changed by the marriage of a French prince and a Florentine princess. Hoping to increase his political power, Francis I of France had made a deal with Pope Clement VII, of the wealthy and powerful Medici family of Florence. The French king's heir, Henry, would marry the Pope's niece, Catherine.

When Catherine de Medici came to France in 1533 for her wedding to Henry II, she brought with her not only her favorite astrologer, Ruggiero, but her favorite perfumer, Renato Bianco, who became famous in France as "René the Florentine." Under the auspices of the crown, René maintained a little shop near the Pont au Change in Paris, not far from Notre Dame Cathedral. He did a lively business not only in alluring perfumes but in lethal poisons. Fine ladies bought deadly chemicals to dispose of rivals and old husbands and lovers, as well as fragrances to attract or hold on to new ones. (René's shop was destroyed during the French Revolution, but some guides can still point to the spot where it stood.)

Catherine de Medici continued to dominate the court and the perfume industry not only during her husband's reign but also during the kingships of her two sons, Charles IX (1560–74) and Henry III (1574–89). Henry III was notoriously effeminate—perhaps his mother was too domineering. His use of perfumes and cosmetics and his behavior brought condemnation from both the clergy and prominent statesmen of the time.

His successor, Henry IV of Navarre, the soldier king, disapproved of perfumes and makeup. His first wife, Marguerite de Valois, however, was a true follower of Catherine de Medici and loved perfumes. Marguerite not only used scents profusely; she is remembered for bringing the new technique of hair bleaching to France. The soldier king's mistress also adored perfumes and insisted on his providing her with large quantities of jewels containing scents.

Catherine de Medici worked her perfume magic on England indirectly. It was to her court that the Queen of Scotland sent her daughter, Mary Stuart, as a very young girl to learn French manners and customs. In 1558, at the age of sixteen, Mary Stuart married the young heir apparent and reigned as queen of Francis II for one year. He died and Mary returned to England and Scotland, where she introduced much of what she learned about fragrances from her mother-in-law.

During the reign of Louis XIII, Henry IV's son, perfume at the French court was revived. But it really blossomed as an industry during the life of the Sun King, Louis XIV (1638–1715). There were perfumed belts, gloves, pomanders, and wigs. There were burning pastes and powders. Louis was called the most sweet-smelling monarch there ever was. In the Sun King's court, it was considered imperative that a different scent be worn each day. Hyacinth, however, remained the perennial favorite, although orange-flower water was popular both in France and England. Louis planted over one thousand orange trees around his palaces and perfumes instead of water played in Paris fountains on festive occasions.

Louis's great-grandson, Louis XV (1710–1774), lacked the judgment and charisma of his great-grandfather, but he inherited his love of scents. In fact, his court was known as *La Cour Parfumée* because a different scent was used each day. His two mistresses' love of fragrance contributed to the depletion of the treasury. Madame de Pompadour spent 580,000 francs annually on scents alone, an enormous sum even in those days. Her two favorites were Eau de Portugal and Huile de Venus. His other mistress, Madame Du Barry, was equally enamored of perfume.

The soldiers returning from the Seven Years' War (1756–63) brought back with them an alcoholic perfume supposedly invented by Jean Paul Feminis, a Milanese perfumer, in 1690. He was said to have obtained the formula for the queen of Hungary's water from a convent in Florence and then modified and improved it with local oils such as bergamot, lemon, and orange. Feminis reportedly gave the formula to his nephew, Jean Antoine Farina, and the latter settled in Cologne, where he called his product "eau de cologne"—still popular today.

Louis XVI (1754–93), who followed his hedonistic grandfather Louis XV, tried to institute reforms, but he was foiled by the extravagances of his own wife, Marie Antoinette, a lover of scents if there ever was one. She popularized sachets made from dried rose petals, sandalwood, cloves, coriander, and lavender. Ironically, her favorite perfume was called "A la Ruine des Fleurs."

Louis and Antoinette lost their heads, but evidently perfumes were valued above royalty. The queen's personal perfumer was Faregeon, who had written a text entitled *The Art of the Perfumer*, a standard work for many years.

Though the queen's head rolled along with the others, Faregeon's remained attached. He became personal perfumer to one of Louis's successors, Napoleon.

Napoleon, the little general who took over the country in 1804, used several bottles of cologne a day—sixty a month. His favorite soap was Brown Winsor, perfumed with a blend of bergamot, caraway, cassia, cedarwood, clove, lavender, petitgrain, rosemary, and thyme, with storax and castoreum as fixatives. His wife, the Empress Josephine, encouraged the use of soap in France. She bathed with it, as well as with rose water mixed with brandy.

Of course, Napoleon divorced Josephine for political reasons, but it is recorded in history that he liked "light scents" and she insisted on wearing musk oil, which he couldn't stand. Josephine, who had thousands of varieties of rosebushes in her garden at Malmaison, decided to make her mark inside the house as well as outside. Before she departed, she sprayed her favorite scents around the palace to the point of saturation in the hope that the Emperor would not be able to forget her.

While France was having its political and fragrant historical upheavals, England had embarked on a similar course. Both Henry VIII and his daughter Elizabeth were amateur dabblers in pharmacy. They made scents and pomades in their castles.

During the reign of the Virgin Queen (1558–1603), the perfume industry was really established. Fragrances had been valued commodities since the Crusades, but they had not been made on the island. Then, in 1573, Edward de Vere, the Earl of Oxford, brought back from Italy, as a gift for the Queen, a scented cloak, gloves, and other fragrant articles of leather.

Elizabeth was so delighted with the presents that she began encouraging her female subjects to cultivate gardens and learn how to blend floral essences with other aromatic materials in the making of "scented waters," pomanders, and sachets for the household. The women followed her suggestions in a manner suitable to their social positions. A woman presided over others or did her own distillations of scents in a "stylling house"—a distilling room—or in the family kitchen.

Musk and civet formed a part of almost every British perfume. Shakespeare, who favored orange-water cologne for himself, conceded that if an amorous swain was to please and awe his damsel, he had to rub civet into his body.

The place in London where the best perfumed products could be purchased in those days was Buckelsbury Street, and by the end of Elizabeth's reign Englishwomen were very Italianate, using quantities of imported cosmetics and scents as well as the fragrances they distilled themselves.

Even after Elizabeth's death, in the seventeenth century, personal cleanliness was still considered quaint. The only really clean people were said to be the Puritans and the Quakers. Soap was first manufactured in England in 1641, but because of harassment from the government through taxes and restrictions the industry was slow to grow.

In England the ascension of the dour Cromwell in 1649 all but destroyed the perfume industry along with the monarchy, but by the time Charles II, the merry monarch, climbed onto the throne in 1660, there were still enough artisans left to revive it. Powders and patches began to make their appearance, and the greatest perfumer of the day was Charles Lilly. He left an excellent summary of the perfumer's art in a book which was not published until several decades after his death.

Aside from their alluring qualities and their ability to counteract noxious odors, certain scents were credited with remedial powers. During the Great Plague of 1665, quantities of aromatic substances were burned in homes and pomanders were carried on the person to ward off the dread disease. The court physician to Charles I, Dr. Thomas Clayton, created an aromatic potion which was used to fight the plague.

Dutch physicians carried walking sticks with hollow handles as receptacles for camphor, musk, or other pungent scents, which they held to their noses to ward off smells and infections from patients.

When a prisoner was condemned to death by hanging in London's Old Bailey, flowers were spread about so their scent could guard the judges against jail fever, and for centuries judges carried a nosegay to guard them against both disease and the smell of the defendants. Flowers are still strewn about the Guildhall today on state occasions.

Fumigation was a rite of purification and strong-smelling antidemoniac remedies were used for centuries. Both bad and good fragrances were thought to have the power to ward off evil. Camphor and garlic were worn as protection.

Fragrances were touted as miraculous by Giuseppe Balsamo, known as Count Cagliostro. This Italian court alchemist toured Europe with his wife and peddled sweet-smelling love philters and perfumes that were advertised as being capable of making ugly women beautiful, a claim not too different from those of the cosmetic promoters today. During Louis XV's stay in France his last mistress, Madame Du Barry, paid him a large sum of money for a potion to preserve her youth and beauty.

One person who credited perfume with almost incredible power was King George III (1738–1820). When thirty-six years old he decreed: "Whether rank or professional degree—whether virgins, maid or widows, that shall from and after this act imposed upon, seduce and betray into matrimony any of his majesty's subjects by the use of scents, potions, cosmetics, washes, artificial teeth, false hair, spanish wool (rouge), iron stays, hoops, high heeled shoes, or bolstered hips, shall incur the penalty of the law now in force against witchcraft and like misdemeanors and that the marriage, upon conviction, shall stand null and void." Despite his efforts, the lingerie shops continued to display scented underwear, and perfumes were used in washing.

George Washington, who proved to be persona non grata as far as George III was concerned, and the pirate Captain Kidd had something in common. They both liked a scent created by a combination of musk, orange blossoms, bergamot, lemon, and twenty-three other ingredients. This concoction is still being sold in a New York drugstore, Caswell-Massey, the same pharmacy it is claimed that served the first American president.

The history of perfume is filled with fanciful tales, intrigue, and romances. The Duke of Tuscany, for instance, came into possession of the first jasmine plants grown in Italy. His gardener, unbeknown to the Duke, gave some plants to his mistress. She grew them and was so successful at it that she made a fortune and was able to marry the Duke's gardener and keep him in fine style.

By the end of the nineteenth century, perfumery had changed from an art to an industry, and in the next two decades new chemical aromatics became available.

Outstanding individuals like Coty and Guerlain were able to create a single new perfume every five or ten years. Today firms like International Flavors and Fragrances and Proprietary Perfumes, Ltd., create about ten daily.

The most romantic of all the early twentieth-century perfumers—or at least the one with the best copywriter—was Jacques Guerlain. He is said to have observed that when a woman is given a fine perfume the giver is saying, in effect, "While I do not till the fields, I am able through enterprise and ingenuity to bring you the essence of their blossoms and thereby I liken you to the most noble of nature's creations." On his way home from work on a summer evening in 1911, he supposedly paused on a bridge over the River Seine in Paris. It was dusk, when "the sky has lost the sun but not yet found the stars." There was a hush in the city, the interlude between the bustle of the day and the gaiety and romance of the night—that fleeting instant when the elements seem to be conspiring to say something tender. It is said that Monsieur Guerlain, having a premonition of the holocaust about to envelop Europe, wanted to memorialize the peacefulness of that evening for future generations. He succeeded so well when he created the perfume L'Heure Bleu that it has remained a continuous best-seller for more than half a century.

It is said that Guerlain named another perfume, Mitsouko, for a Japanese girl who fell in love with a British naval observer during the Russo-Japanese war; the lovers were separated but whenever the naval officer's heart turned eastward, whenever he encountered the mysterious, heady scents of the Orient, he thought of his lost love. Guerlain brought out Mitsouko in 1921, and it is still selling today.

He created Vol de Nuit (Night Flight) in 1933, in celebration of a book by that name written by Antoine de Saint-Exupéry, the French aviator-poet. It tells of the brave men who pioneered the airmail routes across South America over the Andes, and "the brave women who awaited them." Monsieur Guerlain said Vol de Nuit smelled of "excitement and adventure."

Guerlain's products are still popular, along with those of his contemporaries, well-known couturiers who persuaded their wealthy patrons to sample house fragrances. Among them were Chanel, Lanvin, Paquin, Patou, Worth, Lelong, Nina Ricci, and Molyneux. Then came the manufacturers of furs such as Revillon and Weil and with them newer designer-perfumers like Balenciaga, Balmain, Pierre Cardin, Carven, Christian Dior, Jacques Fath, Givenchy, Madame Grès, Pucci, Schiaparelli, and Yves Saint Laurent.

In Spain, perfumers who became well established in the 1920s and 1930s remain successful. They include Dana, Myrugia, and Puig.

A new perfume can still be introduced today by a designer, although, according to Halston, who introduced one in the early 1970s, it costs about one million dollars to develop a new scent. Jovan, a Chicago company founded in 1969, spends three million a year on advertising its own perfumes.

There are thirty-three large perfume manufacturers in the United States today, of which Avon, Revlon, Charles of the Ritz, and Estée Lauder are the biggest.

Unlike the ancient masters with their restricted palettes, a perfumer today has about five thousand raw materials from which to choose, but our use of fragrances is quite similar to that of our ancestors. The woman who wears perfume behind her ears or the man who chooses a pleasant after-shave lotion is behaving in much the same way as an ancient Egyptian or an Elizabethan who used fragrance to lure the opposite sex.

7 The Scent Manipulators

DURING WORLD WAR I, the major defense weapon of the skunk—butyl mercaptan—was used by the United States Army as a camouflage for poison gas. When German intelligence became aware of this, their soldiers were warned to put on their gas masks whenever they smelled a skunk. Since wearing these cumbersome protection devices severely reduced their fighting efficiency, the Americans eventually released the skunk odor alone without the lethal gas. Thus, the German forces were hobbled with gas masks while they themselves could maneuver without the fear of backwinded poison gas.

More than three decades later, another battle was initiated by a powerful force of mind manipulators—a battle whose objective was to induce the consumer to choose one brand of merchandise over another. In the forefront of this battle was an experiment in market research which involved three identical batches of nylon stockings. One batch was scented with a fruit fragrance, another with a floral fragrance, and the third was left with its own synthetic odor. Dr. Donald Laird of Colgate University, who conducted the experiment, then observed women shoppers touching and inspecting the stockings. They generally chose the floral-scented stockings as being "softer and more durable." Only 8.5 percent of the shoppers chose the unscented pairs as being most desirable. Of course, the single variable among the three batches was the smell.

That experiment paved the way for a whole new field of mind manipulation. Now we continually are lured, without our conscious knowledge, into buying products because of the way they smell. The fact is that less than 20 percent of all fragrance presently employed is in toiletries and perfumes while 80 percent is used to scent other things, from laundry detergents and furniture polish to used cars and glue factories. Included among the deliberately perfumed items are greeting cards, nail polish, underwear, fountain pens, paints, stationery, window cleaners, medicines, tea, and tobacco.

More than $500 million a year is spent making such products smell pleasant. Today, according to many marketing experts, how a product smells is more important than how well it does its task.

The original reason for adding fragrance to soaps, shampoos, detergents, and plastics was to cover foul odors inherent in the materials. Then, in 1966, Procter and Gamble decided to use a lemon scent in the detergent Joy to give consumers an idea of the cleanser's "natural" grease-cutting cleaning ability. The idea caught on, and the use of fragrance in household products has grown more than 15 percent annually. Smell has become an important psychological signal that we are doing something right when we use a particular product.

In fact, selling by smell has become a big industry. The 3M Company has had spectacular success with its scratch-and-sniff scents. These are microfragrances encapsulated in millions of tiny plastic bubbles—more than fifty million to a square inch of paper. When you scratch or rub the paper, the scent is released.

The Scent Manipulators

Chesebrough-Pond's division of Cutex used scratch-and-sniff labels to introduce its new "Fresh Scented Things." Within ten days, the entire initial production was sold out.

Fleischman's Distilling Corporation was able to hold on to its over-the-bar business in martinis and whiskey sours even though it raised its liquor price three times in two years. Magazine scratch-and-sniff advertisements which smell like martinis and whiskey sours received the credit.

Fragrances are used to make us buy in more subtle ways. For instance, carpeting has a foul odor when first produced. Manufacturers tried to change the image of carpeting by applying such names as "cinnamon" and "rose" to their products, but it didn't help much until the actual scents were added in the manufacturing process.

It took a panel of scent experts a long time to determine a "jaguar" smell which could be sprayed inside a Chevrolet to give it an expensive aura, but they did it. An easier task was to create a general new-car smell—a blending of oil, leather, and metal scents—which is now sprayed on used cars to give the subconscious impression of newness to the potential purchaser. On the other hand, unscrupulous antique dealers use a mixture of musty smells to make new things "old" for the unwary buyer.

Plastic shoes which closely resemble leather weren't doing well in the marketplace until they were sprayed with the odor of leather.

Students in Macomb, Illinois, proved in an experiment that people buy twice as much popcorn in a movie theater when the odor of buttered popcorn is wafted through the air, even though no popcorn is in sight. The synthetic odor of strawberries has been successfully used to make customers buy frozen strawberries on sale in supermarkets. We have all noticed the delicious scents of a bakery, but few of us realize that this scent is often deliberately vented out to the street, where passersby can catch a whiff.

The British researcher Juanita Byrne-Quinn, who looks like a kindly schoolteacher, is one of the world's leading marketing experts in the field of fragrances. She works for Proprietary Perfumes, Ltd., one of the largest manufacturers of perfumes worldwide, and it is her job to determine which smell will sell a product best.

"First," she explained, "fragrance in itself evokes a variable response in the user. It can be as simple as 'nice' or 'not nice' or something in between, and the response can vary with the kind of individual, his experience, and his expectation of the product.

"On the other hand, fragrance may be acting only as an underlying ingredient in the formulation—for example, masking a bad base odor yet not itself creating a noticeable smell. In this case, the user may be completely indifferent to the smell of the product until the fragrance effect is withdrawn.

"Fragrance, however, not only has a function of its own; it is also a medium for a message. The messages may be directed to the user about the product or they may be in the product to inform other persons about the user's personality or self-image.

"Perfume is used in many ways throughout the world," she continued. "Consumers will judge, in part, from the smell of the product whether it is likely to be efficient, whether it will care for or harm the skin, or whether it is suitable for children."

Moreover, fragrances can change during the usage cycle of a product. So before Miss Byrne-Quinn begins any market research related to fragrance, she has to consider precisely which aspect she is going to deal with

—the consumer, the fragrance, the message, or a combination of all these aspects. The art of adding the right perfume to the product is equally the art of choosing the right tools for the job.

Miss Byrne-Quinn travels widely, testing impressions based on fragrances. In Germany housewives were asked to use laundry products for three weeks and then make an overall evaluation of their total effectiveness. The only variable in the products was, of course, the scent. But the results of the questions dealing with other attributes encouraged Proprietary Perfumes, Ltd., to examine these messages closely. Exploratory research showed that the consumer response to odor in, for example, toilet soaps is multidimensional, involving four sensory modalities—sight, touch, taste, and smell.

"Take the word 'freshness,' for example," Miss Byrne-Quinn said. "There is the freshness of newly baked bread and the freshness of a rose on a summer morning. There is the freshness of a walk along the beach with a fresh wind blowing. There is the freshness of newly laundered linen. These sensations all use the same word and depend on the sense of smell. Yet, for their definition they must use the other senses. The freshness of bread is closely allied with taste and the freshness of linen with touch. The freshness of a rose on a summer morning is very different from the freshness of a walk along the beach."

This means, Miss Byrne-Quinn said, that perfume is not just a gimmick used in advertising but, like advertising, is also capable of influencing the perception of the product. "Perception depends not only on the senses but also on the previous experiences and present awareness of the one perceiving. We know, not only intuitively but also from systematic research, that housewives judge their washing predominantly through sight and only marginally by touch and smell. And yet, their final perception of 'cleanness' can be measurably influenced by smell, as we proved in Germany."

Such product fragrances must be chosen according to task rather than culture. "Women using an automatic washing machine in Brazil and in Paris want the same perfume, but women in those countries doing their wash by hand in cold water want another kind of perfume. The women with the washing machine will want the clothes to smell fresh after they take them out of the machine. The women doing their own washing want them to smell nice while they are washing them."

Culture, on the other hand, does influence the choice of personal perfumes. "We can see that personal perfume is primarily concerned with personality and has few if any connotations of task. Toothpaste and toilet soap closely involve the consumer's personality. A washing powder does not. It is basically functional and a floor cleaner is almost entirely so. These relationships give us some idea of the direction we must look for in terms of motivation. For example, when dealing with a personal perfume, we concentrate almost entirely on the personal motivation of the consumer. When we look at personal cleaning products (such as shampoos), we must be concerned with both personal and task motivations. With a product such as a floor cleaner, we concentrate almost exclusively on what motivates the consumer to the task."

On the social side, a perfume may help to reinforce the wearer's confidence among her reference group, or confirm her in her peer group, or move her toward her aspirational group. Biopsychologically, perfume may be used to help to emphasize individuality, to achieve supremacy, to establish a concept of self, and to attract sexually.

Perfume is closely allied with both personality and nationality. The British researcher says research has shown that in the United Kingdom, where there is a puritanical streak, it is not considered nice to use too much perfume, so Britons try to use a smooth, soothing scent. They use more bath additives, hair sprays, and talcum powder than perfume.

Frenchwomen, on the other hand, scorn talcum powder. Toilet waters and colognes are used in large containers in that country. The French use perfume to demonstrate good taste. They are careful not to offend, so they don't use strong perfumes.

Americans wear perfume to be accepted within the group. They carry more than one perfume and change it according to mood. Americans like so-called "stinker" perfumes. Estée Lauder made a fortune by doubling the powers of all her fragrances.

In Germany perfume is used to display not only good taste but also wealth. The more expensive the perfume, the greater the aura of personal wealth it conveys.

According to Miss Byrne-Quinn, it is a mistake to think that men don't like perfume. They may, however, prefer to perfume themselves with soaps and deodorants instead of colognes.

The commercial manipulation of our minds by the use of fragrances is a combination of modern research and ancient techniques. One of the most fascinating aspects concerns the people who actually create the scents —the perfumers. Instead of the painter's colors or the musician's notes, they blend fragrances to convey a message—whether it is sensuousness for an evening perfume or crispness and cleanness for a laundry detergent.

The truly creative artists of the industry are called

"noses." There are only about twenty in the entire world who are considered great. Their talent is inborn, although it may take many years for them to perfect it.

Classic perfumes contained about thirty ingredients, but today's new fragrances may contain two hundred to three hundred substances. "Noses" not only can distinguish between the scents of tangerine and orange but can detect the individual ingredients in a mix of a hundred or more. They can discern almost the precise amount of the various substances which contributed to the blend. They are the olfactory counterparts of wine tasters, but their ability far surpasses the relatively simple tasks of tasting.

"Noses" know what happens to various scents when they are blended. They have committed more than two thousand fragrances to memory. Their skills far surpass the capabilities of any "smelling" or chemical analytical device yet invented.

Perhaps one of the leading "noses" in the world is Ernest Shiftan, vice-president of International Flavors and Fragrances. He first became interested in perfumes as a child. "I was twelve years old and I discovered how beautiful my mother's perfume smelled. I decided to study chemistry to see how perfumes were made."

Shiftan attended schools in Vienna and then joined the I. B. Farben Chemical Company. He later established his own perfume business but had to flee the Nazis. He went to France, where he again established his own business and again had to escape when the Germans marched into that country.

He came to the United States in 1940 and joined the firm which eventually became International Flavors and Fragrances (IFF). The United States Army Intelligence interviewed him shortly after his arrival and asked him if it would be possible to train people to recognize by smell the nationality of a soldier who had captured and blindfolded them. "I told them if the soldier was eating his regular foods, then it would be possible to recognize his origin if you were trained to do so."

Shiftan, who is still in love with fragrances after his many years in the field, said that there is no school for perfumers. He is training perfumers for his company, which has about twenty-five perfumers in the United States and an equal number in foreign branches. Some of them specialize in soaps and detergents, some in industrial perfumes, and some in toiletries and high-quality personal perfumes.

"The perfumer uses his sense of smell just as a musician uses notes. A musician knows exactly what tune you play on the piano. A perfumer knows what is in a perfume. The great difference is that in twenty years, you can sit down at the piano and play a musician's score but you cannot reproduce a perfumer's perfume."

Shiftan said there are vintage years for the natural raw materials of perfume, as there are for wines. The quality of the perfume depends on the amount of rain and the time that its natural ingredients were picked and shipped. The perfume industry today is suffering from changes in the world, he pointed out. "It is becoming more and more difficult to get good natural materials. The area of land on which the natural products are planted is getting smaller and smaller. The cost of the land gets higher because of population growth. Factories are built instead of roses being planted. We have to go to other countries to plant flowers, where the land is not so expensive. And then the workers would rather work in a factory than pick flowers.

"The crops are different each year. Sometimes they

are very good, sometimes not so good. You can tell the difference, just like wine. Take lavender oils, for instance. There are dozens of them. The higher the plants are grown, the better the oil."

There are some five thousand natural and synthetic materials from which to choose, and the perfumers must have committed most of their odors to memory.

"A perfumer must not only recognize many odors," Shiftan said, "and have a good sense of smell and a great memory; he must be creative. We have several people here [IFF, New York] and in foreign branches who want to be perfumers. We test them to see if they have a keen sense of smell. I show them certain products which I know they have smelled during their life—for instance, vanilla, fruits, ham. They not only have to have a sense of smell, they must have a memory of smells to recognize them. Then we have to test for creativity. You can have an excellent sense of smell and not be creative.

"A person who wants to be a perfumer and seems to have the ability will go first to IFF's odor-control department, where he or she comes into contact with all the company's aromas. They will have to stay there one to three years and get to know anywhere from five hundred to three thousand fragrances.

"Then the would-be perfumer comes here, to our training laboratory, where they have to try and make mixtures. They're also given certain things—sometimes difficult things—to imitate. This way, they learn how other mixtures are made.

"The training takes about five years, sometimes as long as ten. If the person really has great talent, then at the end of it he or she is considered a perfumer.

"A number of perfumers have a degree in chemistry, but that is not considered necessary. The perfumers are well paid while they are being trained, and a professional nose makes anywhere from $20,000 to $100,000 a year."

Of his own ability to recognize perfumes Shiftan said, "I can pick up a scent and recognize what's in it. When I go to a party, people who know me will say, 'Guess what my wife's wearing.' But perfumes smell differently on different women. Sometimes I can't even recognize my own perfumes on them.

"Some women like people to ask them what perfume they are wearing. Other women don't want to be recognized as being perfumed. They want people to believe that the perfume is part of their own odor. These people use less perfume. The Frenchwomen are this way. They just want to enhance their own odor."

Shiftan said a perfumer can smoke. "I know a French perfumer who smokes a pipe from morning till night."

He, himself, likes no odors at all when he sleeps. "I am smelling all day long, so at night I like to give my nose a rest."

His favorite smells, of course, are perfumes—particularly woodsy perfumes. He hates cheap perfumes. He also hates a popular American ammonia cleanser which he thinks smells like a French pissoir.

"Perfumers are used in different ways," he concluded. "One may be for daily routine work while the other is for creating new compositions. New creations sometimes start with a brainstorm, but the process can also be very slow and drawn-out and take from two to five years.

"The genius in perfumery must have a natural feeling for femininity and masculinity. He or she must have discerning taste and a creative streak, with an instinct for odor harmony and the intelligence to choose the right product to achieve that harmony."

What would Shiftan like to perfume in the future?

"Things that smell bad. I would like to make them smell pleasant—hospitals, for instance. There is a natural instinct to improve the odors around your body and the odors in the air. We may have gone a bit too far with deodorants. Some of them are too pungent. The odor of a clean body can be very pleasant."

Hugh Watkins is a young perfumer at PPL's plant in New Jersey. He works in a room filled with shelves of bottles. The combined scent is overpowering when an outsider first walks in, although Watkins claims he does not notice it. And yet he, like Ernest Shiftan, is acutely conscious of smells in the outside world which might be imperceptible to the average person.

When a client comes to him to make a perfume for a product, Watkins has to determine not only the perfume's function as an ingredient and how it will survive in the product, but what the advertising platform for the product will be. "You also have to consider cultural likes and dislikes," he said. "For instance, wintergreen, which Americans like, reminds Europeans of a disinfectant, so they do not like it in shampoos. Nordics like the scent of a pine forest. Latin men like citrus odors.

"When considering a product perfume," Watkins continued, "you have to consider the odor types required. Will the perfume be stable in the product? Will the active ingredients in the product, like chlorine, help or harm it? Then you have to consider the price. How much will you be allowed in terms of cost? You can't scent a cheap scouring powder with an expensive perfume."

Watkins said he has been asked to create perfumes for a VD cream, toilet paper, pigs, oven cleaners, aircraft sanitizers, and, of course, for personal use.

"A woman does well to smell clean," he commented. "She can look and behave wonderfully, but if she doesn't smell nice, it's to no avail. On the other hand, if she dresses and behaves atrociously, it won't matter what perfume she's wearing."

He concluded that perfume is only part of the picture—an important part but still just one factor.

As far as the cosmetics industry is concerned, however, perfumes are becoming the fastest-growing division. Fragrances used to be the smallest segment of the beauty business, but within the past five years, sales of scents have increased by nearly a billion dollars.

A commercial fragrance is basically composed of three things:

1. A solvent.
2. Odorous substances—sometimes as many as three hundred in a single perfume.
3. A fixative. Without a fixative a lasting perfume would be impossible, since ingredients might evaporate at different rates, making the scent smell completely different after a period of time.

There are certain basic ingredients in perfumery. First there are the "essential volatile oils" derived directly from the leaves, flowers, fruits, stems, woods, and roots of plants. Such oils come from around the world—lemongrass, patchouli, and sandalwood from India; petitgrain and rosewood from South America; peppermint from the United States; lavender from Spain; ilang-ilang from Malaya; cloves from Zanzibar; and roses from the south of France, Bulgaria, and North Africa. Gums and resins contain a percentage of volatile oils.

There are four main animal products used in perfumery: ambergris, which is formed in the intestines of the sperm whale; musk, which comes from a small gland near the sex organs of the male musk deer or from the muskrat; civet, a musky substance produced by the civet cat; and castor, a glandular secretion of the beaver. Such animal products are mainly used as fixatives.

Most of the musk being promoted so heartily as a sex attractant in perfumes and cosmetics is synthetic. And with good reason. Hunting expeditions have to be organized to seek new supplies of animal scents. The musk deer of the northern Himalayas browses on the snow line of the Tibetan mountains and is elusive, moving mostly at night. Its musk can only be obtained when it is killed. The substance is then taken by caravan through China to the Burmese border. Musk products from the southern Himalayas go through India. Civet from Abyssinia passes through many hands before it arrives at a London perfumery depot, a brown, greasy mass still packed in the hollow horn of a zebu, the humped domestic ox of northeast Africa.

Because of the expense and difficulty of obtaining natural products, some odorous substances are either derivatives of natural substances or completely developed within chemists' laboratories. Modern chemists are pretty good at imitating nature, and some of their creations are more practical than the originals.

The rose, for instance, is not pure. It consists of many chemical compounds that vary from rose to rose and from season to season. If the essential oil dissolved from a rose were put into a bottle, it would soon evaporate. Its fragrance would alter before it was used up. It would be difficult to scent products such as soap and face cream with it, and no one would be certain what the next bottle of perfume or the next lipstick would smell like. The perfumer has to intervene and give the natural oil stability, persistence, and uniformity, as well as getting from it an economic yield and enabling it to be applied to commercial substances.

PPL's Hugh Watkins maintains that "The perfumer's job is to create or compose fragrances in a form which will be of practical value to the manufacturers of perfume and toilet waters, as well as of branded cosmetic soaps and many other commodities. The cost of essential oils has skyrocketed and may soon make all natural perfumes out of most people's reach. On the other hand, the price of oil, on which many synthetics are based, has also skyrocketed and so have the paper and plastic in which the products may be wrapped."

The perfumer, he continued, has to work with the customer's product base in mind. He has to know which raw materials are suitable for each particular use. For example, in the case of a detergent powder for use in a country with a hot climate where distribution is relatively slow, the perfumer will choose an odor which will not be lost or change too much in character during the lengthy product shelf life. He will also know which materials are best suited for powders in automatic machines, whether the manufacturer wants maximum odor impact when the package is opened or fragrance during the wash. He even knows what will happen to the fragrance during the ironing of vegetable fibers such as cotton.

Both Watkins and Shiftan noted that today's perfumer is not free to dream up a composition like a musician. He is restricted by the availability of raw materials, their costs, and the safety regulations of the U.S. Food and Drug Administration.

In the 1960s, additional restrictions were put on the use of certain raw materials in fragrances which might create a problem of safety, irritation, or toxicity. Then came the tremendous increase in prices of raw materials. The manufacturers of perfumes met the challenge by putting their fragrances into toilet soaps, lotions, and other less concentrated forms.

Nevertheless, the future looks rosy for the perfumers. As humans gather together in cities, as the use of servants disappears altogether, there will be a growing need to make things smell better.

Furthermore, the Food and Drug Administration has ruled that all cosmetic ingredients, with the exception of fragrances, must be listed on the label. The only mysterious thing which will remain will be the perfumes.

Perfumes are also difficult to copy. As Hazel Bishop observed, "They could copy my new lipsticks and cosmetics, but they could not duplicate my fragrances."

Today's perfumes are at least twice as strong as those of the last century. Both men and women smoke and drink a lot, so more perfume is needed to overcome these odors and make an impression on the olfactory sense, which is weakened by these habits. Another reason is that air conditioning removes the aura of perfumes from around the body.

The perfume business is therefore strong, literally and figuratively.

The combined power of advertising and perfume is awesome. Colgate's Irish Spring soap, for instance, was heavily advertised on television and in the print media. A year after its introduction this soap, with a heavy dose of IFF fragrances, had captured 8 percent of the market. Television commercials showed rosy-cheeked Irish citizens with heavy Irish brogues in the invigorating Irish air surrounded by the green Irish countryside. Actually, the scent for Irish Spring was developed and produced in the metropolitan New York area.

IFF's colorful president, Henry Walter, Jr., who rides his bike to work in New York and wears suspenders decorated with skunks, sells smells with vigor. At a meeting of European investors he once took off his shirt and splashed fragrant lotion on his chest, ticking off the qualities of the perfume. While the money men gaped, according to the trade magazine *Institutional Investor*, he told them about the alarmingly high VD rate in Scandinavia, which led the government in that part of the world to distribute free prophylactics. But the effort proved futile at first, Walter said, because women disliked the rubbery smell of the things. "The room became charged with subdued hysteria," according to the trade magazine report, "while Walter thrust his hand into another carton and produced a fistful of IFF's scented solution—free samples to each European."

Half a century ago, perfume was an extravagance used only by a few sophisticated women. Today, it is used not only by most people but also in a growing number of products. It is expected that there will soon be total body perfumes and room perfumes for different occasions and moods.

In the meantime, there is a new pen called a VAP that dispenses fragrances. Introduced in 1975, it consists of a reservoir at the top of a ball-point pen, a highly absorbent filler called "woolex," and a device to regulate the speed of evaporation. The scent of choice is injected into the woolex by means of a dropper. By twisting the top of the pen, the user controls the amount of fragrance that is exposed to the air. Its inventor, A. Oscar Lin, said the pens make it possible for the user to enjoy a breath of fresh, fatigue-lifting pleasure whenever he needs it.

A company called Smell This Shirt was one of several organizations of its kind organized in 1975. Smell This Shirt uses scented T-shirts for commercial messages as well as for fun. For instance, an advertisement on a shirt for Clairol's Sunshine Harvest Shampoo smells like orange peels. The smell power remains for up to fifteen washings.

Other smell shirts include "fish odor" for a bait and tackle company, the odor of diesel fuel for a machinery manufacturer supplier, and apple pie for an Indianapolis television station. One of the biggest sellers was the Pot Shot Shirt, which smells of marijuana.

A company in Miami has introduced Snif-T-Panties, women's underwear scented with a variety of fragrances. Among the scents: banana, rose, popcorn, whiskey, pickle, and pizza. Why a woman would want to smell like a pickle or a pizza remains a mystery.

Men's underwear and socks began being manufactured with built-in deodorants in 1974 and 1975. Since the smell of sweat is primarily due to the action of the bacteria on perspiration, the underwear and socks have built-in bactericides that last through many washings.

Artificial fragrances have long been adding zest to food, from a buttery aroma for instant mashed potatoes to fresh-baked bread smells for bread. One manufacturer tried to put the scent of hot pizza on the wrapper of frozen pizzas, but the FDA prohibited this because the chemicals from the wrapper might migrate into the food.

Sometimes adding scents to products has unexpected results. Manufacturers made millions between 1966 and 1970 by convincing women that females have a genital odor which needs to be covered up by perfumed aerosol products. Fifteen sprays were on the market, including one called "Cupid's Quiver" which came in several scents and flavors, among them mint frappé and honeysuckle. But by 1971 the FDA began getting reports of adverse reactions, ranging from burning and itching to infections. In a few cases there were serious inflammations of the urinary tract. The FDA then announced that no therapeutic advantages have resulted from the use of feminine hygiene deodorants and there was a danger of side effects. The sales declined somewhat, but there is still a wide choice of products to cover up the natural female odor.

The California Newspaper Publishers' Association warned its members about another side effect of scented products. A meat company produced a scented-ink newspaper advertisement that smelled like bacon. Neighborhood dogs went a little crazy and began carrying newspapers off the porches and ripping them to shreds. Then, a substitute material was used in a batch of news ink, causing newspapers to reek of fish. Presumably, that was when the neighborhood cats went crazy.

In spite of such setbacks, selling by smell is a rapidly growing, lucrative business. The 3M Company, for instance, has a burgeoning library of standard smells, and the demand for them is increasing, especially in the educational field, where fragrances are used to aid learning. The 3M Company learning kits have scratch-and-sniff labels whose scents are associated with the written and auditory materials. The smells not only help fix things in the youngsters' minds by association; they act as rewards as well—particularly the ones with the smell of chocolate cake. The kits have been so successful that they are being used to help retarded as well as normal children increase their reading skills.

The Braille Institute of America and the Perkins School for the Blind place scratch-and-sniff labels on braille pages and cards to help students learn. Odors are being used to help teach deaf-mutes to speak and to read.

City children are being given a country experience at the New York Museum of Natural History where IFF supplied a cut-grass smell and an acrid marsh smell for appropriate displays.

The link between odors and learning was proved

at the Research Center for Mental Health at New York University. Students were given a word list to learn. Some lists were given without accompanying odors and some with them. The researchers found that when the odor and words were presented together and were related—the word "cheese," for instance, and the smell of cheese—the students remembered lengthy word lists more easily and retained the memory indefinitely.

You can use a child's natural pleasure in smells not only to help him with his schoolwork but to encourage him to eat more nutritious foods, brush his teeth, and keep himself clean. You can do this by giving him foods and cleaning products with smells that he likes.

Since scratch-and-sniff was introduced in 1967, the use of odors in advertising and promotion has skyrocketed. Darrel Huebner of 3M recalled that one company, Pine Top Lakes, used pine fragrance in its direct-mail appeal to sell real estate. Another company, American Republic Insurance, thinking that people associated mint with money, used a mint-scented dollar in their direct-mail advertising. United Airlines used the sweet smell of oranges to lure travelers to California, and the gas industry used a foul-smelling strip which alerted customers to the scent of leaking gas.

Huebner said that the use of fragrances can bridge the credibility gap between consumer and advertiser. It is effective because such promotion offers the appeal of both sight and smell.

Don't sniff at it. The manipulation of our minds by scent is a multibillion-dollar business and growing all the time.

8 Malodor Maladies

ONE MORNING in the fall of 1975, I entered the elegant lobby of the Hyatt Regency Hotel in Atlanta, Georgia. I was standing in line at the information desk when I smelled a distinctly foul odor. I looked suspiciously at the man in front of me and then at the lady behind me.

I walked through the lobby and still smelled that smell. "Maybe the bathroom plumbing has backed up," I thought to myself. But, when I stood outside the hotel waiting for a cab and the unpleasant odor persisted, I was forced to consider: "Maybe it's me?"

It wasn't! The mystery was soon solved by a front-page article in the *Atlanta Constitution*. That unpleasant odor permeating Atlanta inside and out was from a paper mill in Rome, Georgia, more than forty-five miles away.

Stench is one of the most irritating forms of pollution. Anyone who has lived downwind from a stockyard or a chemical factory knows the problem all too well. But on a hot summer day even clean neighborhoods may develop stinks from rotting garbage and animal droppings. Malodor is the least understood pollution. One major reason is that, unlike human responses to light or sound, olfaction cannot be measured.

Particularly puzzling is how smells mix in the air. A combination of odors may be more or less pleasant and more or less intense than one of them alone.

It has been found in a number of studies that those people who are nervous or who have an underlying physical disorder, such as asthma or heart disease, are most troubled by malodors in the community. But, although most of us are inclined to take pleasant odors for granted —we may even fail to enjoy the invigorating smell of salt air at the beach or the scent of pine in the park—few of us can be inattentive to a stink. Malodors, therefore, can be costly in terms of property values, social relationships, and the enjoyment of life in general. They can be emotionally and physically destructive. Suppose, for instance, you are driving behind a diesel bus. It may not be safe to overtake it, but because of the fumes you become so uncomfortable that you are willing to take a great risk to get out of the polluted airstream.

It may be possible for you to get away from a single source of pollution such as the bus, but it is not as easy to move your place of residence.

Since odors can travel far—as far as the forty-five miles from Rome to Atlanta—they can be very difficult to eliminate, especially if they emanate from the waste products of industrial operations. Pulping plants have digester blow systems which release hydrogen sulfide (rotten-egg smell), ethyl mercaptan (skunk smell), and dimethyl sulfide (decayed-cabbage smell), all of which are offensive to most people, even in minute quantities. The main emissions from the by-product coke ovens in steel mills include not only smoke and dust but also hydrogen sulfide, phenols, and ammonia. Oil refineries give off mercaptans from cracking units, phenols and naphthenic acid from scrubbing-solution storage tanks, and hydrogen sulfide. The smell of decomposing proteins, often associated with rendering plants, is an offensive mixture of hydrogen sulfide and putrescine (which smells putrid), skatole (which has a fecal odor), and butyric acid (which smells like sweat).

Kettle-cooking processes in the manufacture of paint and varnish permit the escape of malodorous hydrocarbons, as do continuous processing ovens for vinyl plastic products. Other industrial processes that are often major sources of odors include acid treatment in thermal-process phosphoric acid plants, wire enameling in magnet manufacture, metal lithographing in can manufacture, and tungsten-filament manufacture.

Odors from all these sources generate a high level of public concern, according to Brian W. Peckham, economist with the National Air Pollution Control Administration. He noted that most people who complain about air pollution cite odors as the problem.

Peckham points out that the problem of malodorous air is nothing new. In the third century B.C., Theophrastus, a pupil of Aristotle, wrote a treatise on stones in which he called attention to the objectionable odors from the combustion of coal. The odors and soot from coal smoke chased at least two English monarchs from London. The first to go was Queen Eleanor in 1257 and the second William III, who moved out some four hundred years later. Queen Elizabeth I, who loved perfumes, could not abide coal smoke, and Edward I became so unhappy over the smoke from London furnaces that he threatened severe penalties for anyone who substituted coal for wood. The ancient notion that diseases came from "bad airs" probably had its origin in primeval air pollutants.

Peckham confirmed that air pollution generated in

one community can cause problems for a distant community. He said that Public Health Service investigators, tracking an interstate odor problem along the Vermont-New York boundary, turned up evidence that under stable atmospheric conditions a wind speed of 5 miles per hour was sufficient to carry the detectable rotten-egg smell of hydrogen sulfide from the International Paper Company mill in Ticonderoga, New York, some 31 miles downwind to Vermont.

In a well-documented case which occurred in December, 1969, an accident at a chemical plant in Carteret, New Jersey, released a cloud of ethyl mercaptan gas which subsequently covered Manhattan from the Hudson to the East River and from City Hall at the bottom of the Island up to 90th Street.

Peckham said that, regardless of the area covered, odors can damage health and deprive people of the use and enjoyment of their property. A survey of court records confirms this. For example, thirty-one homeowners brought suit against the Weyerhaeuser Paper Company to recover damages caused by odors emitted from the company's paper mill in Elkton, Maryland. The variety and extent of the injuries, especially to health, were given in the court testimony. One woman said that the odors kept her from sleep, caused nausea, and on five occasions actually sickened her to the point that she lost her breakfast on the way to get her car out of the garage. Another person attributed frequent chest pains to the odors. One family complained that the odors not only interfered with sleep but also drove away guests and forced the closing of all windows and doors. Apparently the mill odors canceled a number of outdoor barbecues and even obliged one family to retreat to an airtight room in order to get some sleep. Of course, it was charged that the plaintiffs were exaggerating their injuries to win larger awards from the court, but the testimony was convincing enough for the court to allow more than $18,000 to the complainants.

Another case involved a rendering plant in Saddle Brook, New Jersey. Plaintiffs told the court that at various times during the year, especially in the hot summer months or when a strong wind was blowing in the direction of their homes, unbearably foul and noxious odors emanated continuously from the plant. These offensive odors permeated the atmosphere and befouled the homes and clothing of these residents, causing some of them to become ill, producing extreme discomfort, dulling their appetites, spoiling their meals, and interfering with normal social and family relationships. The court ruled in the plaintiffs' favor and issued an injunction against the plant. The injunction was subsequently upheld by a higher court.

Peckham points out that bad smells do not respect neighborhoods, not even those of the rich and powerful. He cites as an example Georgetown, one of the most prestigious addresses in the District of Columbia. Residents there complained frequently about the obnoxious odors drifting up from the Hopfenmaier rendering plant by the Potomac River. On warm days many homeowners, including senators, could not sit out in their elegant gardens or leave the windows of their homes open.

Even public agencies are sometimes troubled by odor problems. Wicomico County, Maryland, offered an 85-acre site to the Maryland Board of Public Works for a mental retardation center. The State was just about to accept the gift with delight when it learned that adjacent to the property was a malodorous chicken-rendering plant. The odors from the plant made the establishment of a center, or any other public or private use, unfeasible.

What can be done about industrial or commercial odors? Such problems sometimes involve the same lack of cleanliness as they do in the home, but more often they result from some process in which unpleasant-smelling vapors are unavoidably added to the air. This is an increasing problem as the population grows and more residential areas merge with industrial districts.

The control of such odors depends upon the disposal of the source, whatever it may be. The smells emanating from a plant must be disposed of either through incineration or adsorption or a combination of both.

The first step to be taken is to insulate such buildings and to provide fans that will produce a slight vacuum inside the building, so that the air enters at the transient openings and goes out at only one opening. Windowless buildings with only artificial light are, of course, best. The walls and the roof should be covered with metal sheets. To provide entrance and exit, particularly for freight cars and trucks, doors should have double locks in which only one door is opened at a time.

There is a variety of odor-dispersing equipment available, from high smokestacks to carbon beds. It is almost always possible to contain an odor with such equipment, so if there is a plant producing stench in one's neighborhood, the appropriate authorities should be contacted. In order to enjoy life and property, citizens have a right to breathe air that is not contaminated by foul odors. According to the definition of the U.S. Department of Health, Education and Welfare, air pollution is "the presence in the air around us of substances put there by the activities of man, in concentrations sufficient to interfere directly or indirectly with his comfort, safety, or health or with the full use and enjoyment of his property."

Since our homes are not hermetically sealed, air pollution may invade them from the outside, but we may also produce our own air pollution within the home. Household odors can usually be controlled by proper cleansing and the intelligent use of disposal methods. Nevertheless, the home is an amalgamation of many smells, not just kitchen odors. There are scents from the laundry, from flowers and house plants, cosmetics and cleaning aids, and the telltale bathroom odors. If someone in the family smokes, there are the stale fumes of tobacco, generally tolerated better by smokers than nonsmokers. Anyone who has been a smoker and then abstained for even a few days is usually surprised to find how the sensitivity to such odors has increased. After a few weeks, the reformed smoker will find the smell of a cigarette that someone else has smoked in the house quite unpleasant—in fact, more unpleasant than the person who has never smoked. In the same manner that a smoker has become desensitized to tobacco odors, a pet owner becomes accustomed to the smell of his or her pet. But when a stranger walks in the door, the presence of a cat or dog is detected by the nose immediately.

In addition to the odors of pets and stale tobacco smoke, there may be woolen fabrics and plastics in the home, some of which give off a strong smell. There may also be fecal material from mice, dead animals in the walls or under the floors, cleaning agents, paint, some lubricants, food dropped on ovens and burners and then repeatedly heated, and fatty foods that have vaporized during cooking and condensed on the walls and furnishings (more than two hundred pounds of grease-laden air is given off each year in the average kitchen).

THE SMELL BOOK

A stranger walking through the door may also smell mildew, decaying building materials and furnishings, and such things as naphthalene, coal tar, Lysol, and the creosote used to prevent damage by insects.

Many of the smells mentioned are not caused by the volatile substances initially present in food or animal excrement, but are produced later by the action of bacteria and mold. The best solution to an odor problem is therefore to find the material that is the source of the odor and remove it as quickly as possible. When it cannot be removed, then its decay or rate of decay can sometimes be prevented or impeded enough to make the condition tolerable.

Most bacteria will not grow without moisture. An electric dehumidifier takes water out of the atmosphere and condenses it. The device should be hooked up to a drain or emptied frequently, so that water doesn't revert to vapor again or serve as a medium for the growth of fungi and bacteria. Calcium chloride, which can be purchased in a hardware store or supermarket, will also absorb atmospheric water. A light bulb constantly burning can dry out a small area but make sure the bulb is not in contact with any material which might catch fire. Washing a moldy area thoroughly with detergent or household bleach can also prevent the growth of spores and bacteria and therefore prevent odors.

Next to cleaning up the sources of odor and drying out the air, the most attention should be given to ventilation. Since odors are the result of airborne molecules, the simplest way to get rid of them is to ventilate the area where they occur. If there is a strong cross-current that circulates the air and the source of the odor has been removed, then the problem is solved. Unfortunately, things are rarely that simple. Many places do not have good ventilation, and some odors persist because they have thoroughly saturated the area or because their source is unknown or not removable.

In most homes, the source of unpleasant odors is localized in the kitchen, the bathroom, or a room that has just been painted. They may not even be considered objectionable in the room of their origin, but when the living room smells like the bathroom or the bedroom like the kitchen, you have a problem and the direction of airflow must be controlled.

Logically, the bathrooms should be in the center of a residential structure with an exhaust fan forcing ventilation upward through the floor cracks. To control odors it may only be necessary to make kitchen and bathroom floors tight by using impermeable floor coverings. The walls may be made tight in the same manner by the use of wallpaper or a thick coat of paint. Even rats' nests in a wall may be made inoffensive by caulking openings and re-covering porous plaster with outdoor waterproofing materials. Windows should be open a little at the top to ensure that these rooms will be colder and air will practically always flow into them from the rest of the house, not in the reverse direction. Even if doors and windows are closed, the average room has twenty changes of air during a twelve-hour period, so nature helps.

If odors persist, there are many techniques to combat them. Flame is the best way to destroy an odor, so a burning candle—particularly a scented candle—may help. But it usually takes too long for any large amount of room air to pass through a candle flame. As many as one hundred ordinary candles would be needed to clear a strong smell in the average living room and this would take twelve hours of continuous burning.

Activated carbon is sold for use in smelly refrigerators and rooms. The object of the carbon is primarily to prevent the transfer of odors from one material—say, fish—to another, such as butter. But the odor of fish cannot get into the carbon unless it is first present in the air, from which some of its molecules will settle into the butter. Odiferous molecules are easily absorbed by fats; butter left uncovered in the refrigerator absorbs the smells of the other foods stored with it. Anyone who has left a smoke-filled room knows that hair retains the smell of smoke for a long time. In both the butter and the hair, it is the fat that traps the odors. In fact, the odor-absorbing ability of hair is the very reason that hair products such as shampoos and setting lotions are pleasantly scented.

When you swallow odiferous foods, such as garlic or onions, along with fats, your breath can be overpowering to others close to you. The fat in our stomach traps the aromas and releases them with the air we expel.

An open box of baking soda has the same advantages and disadvantages as the carbon. Baking soda in solution is also used as a deodorizer for pots, pans, and counters. The best way to prevent the transfer of odors from products within the refrigerator or on shelves in the house is to keep them tightly sealed in a glass container or can, or with plastic wrap.

Certain odors within the home are very difficult to control. One, of course, is the smell of a sickroom, particularly if the patient has cancer or is incontinent. There is a close connection between perception of such odors and the emotions.

For an incontinent person or the patient with cancer, one end of a rubber tube may be placed under the blankets of the bed and the other end connected to a simple suction device, such as a jet pump, which can then be connected to the plumbing. This is the most effective way of counteracting the odor and, if done without producing a sensation of draft, it is comfortable for the patient.

Hospitals, of course, have a depressing smell because of the mixture of disinfectants, sick bodies, and medication. It is easy to see why people send flowers because they not only add color but also cover up the odors with perfume. Incense has been used for centuries in rites of fumigation and purification for the same reasons.

Incense and flowers, used in this way, are examples of deodorants. A deodorant, in the sense of something added to the air that affects our perception of other odors, is a masking agent, an anesthetic, an irritant, or a combination of the three.

Masking involves the overpowering of one smell, usually an unpleasant one, by another smell which is stronger and more pleasant. Weak odors are not perceived in the presence of strong ones; sometimes odors of the same strength can be blended to produce a combination in which one or both of the components is unrecognizable. However, we rather quickly lose awareness of any odor of constant intensity, so there is a tendency for the effect of the masking odor to diminish and eventually to disappear entirely as we become accustomed to it. As soon as the masking odor disappears, our sensitivity to other odors returns to normal; we smell them just as well as before the masking substance was introduced unless the two odors are very much alike.

There are many commercial masking agents. Government researchers have found that the best scent at masking disagreeable odors is a combination of pine, cedar, and sawdust. However, not everyone finds this scent agreeable, for our like or dislike of an odor depends largely on association of the scent with pleasant or unpleasant experiences and impressions of the past. Asked by researchers to pick from among a variety of scents those that added a desirable freshness to the air, a man chose smoked herring and a woman creosote. The man had spent every summer in his childhood playing around smelly fish-wharves and the woman had played in a railroad yard which treated its ties with creosote. Some people associate Lysol with hospital cleanliness and with places which have been disinfected, so they either like it or dislike it on the basis of their associations with these places.

While masking agents cover an unpleasant odor with a stronger, pleasant odor, anesthetics and irritants affect the perception of pleasant odors as much as they do unpleasant ones. To the extent that anything destroys all unpleasant odors, it destroys all odors and flavors whatsoever.

The so-called odorless deodorants are sprays which, although perfumed, do not depend entirely on masking for their principal action. They have an oily base which reduces the ability to smell by a large factor, probably by the formation of a film of oil on the nasal membranes. The combination of this action with the masking odor of the perfume causes the immediate disappearance, to the observer who has inhaled the vapor, of weaker odors initially present. Only the perfume of the deodorant is perceived, but this effect disappears after fifteen to twenty minutes.

When anesthetics such as glyoxal are used to deodorize, there is little or no effect until after fifteen to twenty minutes of exposure. Then the distance at which odors can be recognized begins to shorten and eventually the olfactory system is "insensitive." Even though the anesthetic may be removed from the air, the numbing effect lasts for up to an hour, probably because the anesthetic has been dissolved on the outer tissues of the tract and gradually diffuses to the more deeply seated nerve centers. Usually, after the hour, recovery is complete and the perception of odor is normal.

Chlorine, ammonia, and formaldehyde are both irritants and anesthetics. Turpentine combines a masking effect with irritation. Irritants differ from anesthetics in that the loss of ability to perceive odors begins to be noticeable with an irritant only after half an hour or more, and even when the exposure to the irritant is discontinued, it is several hours before maximum effect is reached. The irritation of such chemicals may continue for days. It has been compared to sunburn, which may go unnoticed during the actual exposure but the effects of which develop and subside with painful slowness. The lasting effect of an irritation has also been compared to a scab formed by a burn on the hand. Just as the scab desensitizes the area it covers to perception of touch or temperature, the effect of irritants and particularly of ozone, the irritant most frequently used for deodorization, is to desensitize the organ of scent. The effect requires several hours to develop fully, but it may last for weeks.

Ozone, even in small concentrations, is dangerous. The ionizers promoted several decades ago in Germany and Russia were merely ozone producers which made people lose their awareness of smells.

Chemicals called malodor counteractants were discovered in the early 1970s by Dr. Alfred A. Schleppnik of Monsanto Flavor/Essence, Inc., in Montvale, New Jersey. Little was done with them commercially for the first few years after discovery because the theory about why they worked was so contrary to established theories of olfaction. However, in the mid-1970s, with the growing need for deodorization, the chemical industry took a new look at Dr. Schleppnik's discovery.

Conventional deodorizers and air fresheners act by flooding the olfactory receptors with a large number of molecules, creating a strong odor that overcomes or masks the malodor. In the process, a much higher total odor level is produced. In contrast, very small quantities of malodor counteractants appear to react with the specific receptor sites involved in smelling such bad odors as perspiration, rancid foods, and amines. Unlike current deodorants which act well against bathroom odors but are ineffective against tobacco smoke or which counteract kitchen odors but are ineffective against the smells of pets, these new counteractants function against most bad odors. They will not affect good smells, such as those of brewing coffee or of roses, but unfortunately they will eliminate the desirable odor of good smelly cheeses.

The net effect of the counteractants is that the olfactory nerves do not perceive the malodor, and there is an apparent lessening of the total odor level.

Test marketing of consumer products containing counteractants, encapsulated in a slow-release form in boxboard, plastic tiles, or paper, has already begun. They may be used in home air fresheners and in such products as depilatories, shampoos, cosmetics, soaps, home permanents, underarm deodorants, douches, pet litter boxes, and industrial products. So far, no adverse side effects associated with prolonged or concentrated exposure to counteractants have been reported.

Some other scientists are skeptical about Dr. Schleppnik's theories of how counteractants work, but whether or not they prove to be ideal deodorants, the more that is understood about how we smell odors and how odors are produced, the better we can counteract foul smells inside and outside our homes.

No discussion of bad odors would be complete without pointing out that sometimes bad odors are good. A great many lives have been saved by bad odors which warn against fire or bad food or toxic gas.

Brown University's Professor Trygg Engen evaluated the possibilities of using odor and taste aversions to inhibit ingestion of harmful substances by young children. But he found, as others had before him, that although children can discriminate between odors they show very little aversion to any odor when they are under the age of five. Thinking on the same lines, the 3M Company has created a kit for school-age children which contains strips of paper encapsulated with "dangerous smells." The odors used are those of harmful plants, gases, and liquids.

Smell tells us all about the chemical nature of things. Harmful things usually smell bad, but not always. Good things usually smell good, but not always. Some things smell good at one time and bad at others. Have you ever smelled food cooking when you were nauseated?

Furthermore, smells which may be pleasant or appropriate for one place may be unpleasant or inappropriate for another. Would you want to eat in a restaurant that smells like a dentist's office? On the other hand, would you let a dentist work on your mouth if his office smelled like a restaurant?

Our principal objective in attempting to get rid of an odor is most often to avoid giving offense to others. We do not want to be embarrassed. Bad odors make us uncomfortable—even sick. We are constantly trying to get rid of them, especially in our compulsively deodorized and sanitized culture. But with every breath we take, with every mouthful of food we eat, our noses are analyzing safety. No method of analysis which man has invented is capable of distinguishing and correctly identifying so large a variety of chemical substances as this single operation—sniffing. Some of the information acquired through olfaction is most pleasurable and satisfying; some is woefully distressing; and some consists of danger signals that may save our lives.

9 The Human Use of Common Scents

WE CAN COMMUNICATE with smells. We can get a message across more effectively with scents than with words or gestures.

The sense of smell is a great gift. It can be used to improve our sex lives, make mundane chores more enjoyable, create beauty, bring back memories, and enhance our learning ability. Unfortunately, none of us use our brain's olfactory ability to its full capacity. In fact, we are taught to push its messages into our subconscious—to ignore them.

There are odors all around us, on us, and in us. There is not a moment in our lives when odor does not influence behavior unless the sense of smell has been lost. Appetizing odors make us salivate, sexy scents arouse us, and unpleasant smells make us feel sick. Rarely do sights or sounds induce such strong reactions because olfactory sensations are primitive and closely linked with emotions.

There is a powerful biofeedback phenomenon concerning our own odors. If we know we smell good, we feel confident about ourselves. On the other hand, we can use personal fragrances to convey messages to others. We can show our supremacy, our individuality, our interest by the way we smell.

Consciously or unconsciously, we permeate our homes with our individual and family scents just as other animals do. Our unassailable feeling of territorial ownership comes from odors even more than from the sight of our household furnishings.

We are luckier than the rest of the world's creatures. They are confined to using the smells which they themselves produce within their own bodies while we can use all sorts of odors obtained externally. To serve our purposes we can pick and choose at will among thousands of scents.

Men and women have always used smells to try to influence their fellow humans, even though it is impossible to know exactly how an individual will respond to a particular smell. When psychologists investigated the link between odors and memories, they found that a single floral scent reminded one subject of a crowded elevator, another of a funeral, another of an old boyfriend who wore too much after-shave lotion, and still another of pollen and hay fever.

Nevertheless, scientific investigations have shown that there are common preferences among sexes, cultures, age groups, and personality types. We can influence others by knowing their preferences and when and how to use their preferred scents.

The Japanese, for instance, perfume almost everything they use in daily life. They even play games with friends and family that involve identifying smells.

To the Anglo-Saxon even to smell one's food or wine in public is an uncouth act, and yet the bouquet of wine and the taste of food both depend heavily on the sense of smell. Anglo-Saxons use subtle scents to perfume themselves; they do not like to be "obvious."

Orientals like the root extract valerian, which is de-

The Human Use of Common Scents

tested by most Europeans. The Japanese like camphor, a bark extract, and borneol, which smells peppery. Camphor is said to keep away the worms that destroy bamboo. Camphor is liked by American introverts, it has been discovered, and disliked by extraverts.

Asafetida or "Devil's Dung" has a fecal smell and is prized on the borders of Asia as a condiment for food. Americans hate it but then, on the other hand, Asians can't understand our love for smelly cheeses.

Northern Europeans prefer heavier fragrances for use in their cold climes, while Mediterraneans like sophisticated floral smells, probably because they love being surrounded by flowers. Orientals appreciate heavy, spicy, animal perfumes.

Most people, of course, like flower and fruit smells, and almost all are revolted by such bad smells as those of rotten eggs, rotted fish, and stuffed drains. It is safe to say there is broad agreement in the human race about what does and does not smell pleasant. That is why so many perfumes smell of roses and jasmine and musk.

Some people are much more affected by smells than others. Sensitive individuals at a party may be attracted by the fragrance of the flowers on the coffee table, while other more pragmatic guests may be looking for the drinks.

Generally, humans like the smell of what is good for them and dislike the smell of what is bad. However, not everybody likes cod-liver oil, and a chocolate cake may smell delicious to a diabetic.

Investigations, particularly those of the British researcher R. W. Moncrieff, have shown that there are certain determinants of olfactory preference. The first in importance is age, followed by sex, then temperament, and finally intelligence.

It has been demonstrated repeatedly that before the age of five, most children think that nothing smells really bad, not even feces, sweat, or amyl acetate. By the time they reach their fifth birthday, however, they have adopted the attitudes that society wants them to have toward such smells.

Young children are closer to the animals, and their preferences for odors are based upon bodily requirements more than are those of adults. Children do not like flower smells as much as adults, for instance, but they do favor fruit and food smells. Adults have usually developed an aesthetic sense and, with it, the ability to appreciate the beauty of certain ethereal scents. The more intelligent children are, the more their aroma preferences seem to mimic those of adults. For some reason not yet clear, youngsters universally dislike oily smells.

Under the age of eight, according to Moncrieff, both boys and girls show a strong preference for strawberry essence. It is their favorite. The boys like vanilla next, and the girls' second choice is almond essence.

In the eight- to fourteen-year age group, boys like or tolerate the smell of orange blossoms better than girls do, and young boys show a marked liking for musk lactone, which has sexual associations. The girls in this age group still like almonds but add an inexplicable liking for the tarlike smell of naphthalene.

In the fifteen- to nineteen-year age group, the differences become more marked between the sexes. There is a sudden preference for lavender, although it is preferred more by young men than young women. Lavender seems to carry a hint of good housekeeping because it is used in many household products. Why the males in this age group like it better than the females is a mystery.

Young men like vanilla just as much as the boys do, but the young women are not so fond of it as they once were. They still like naphthalene better than men, and this is consistent through all the age groups.

In both sexes, the peak of either liking or disliking an odor occurs about the age of ten. Strawberry, vanilla, musk, and orange blossoms all show peaks of liking at this age. A strong dislike of chlorophyll occurs at ten years.

Generally, the biggest changes in olfactory preferences take place within the first twenty years of life, although there are exceptions. Both sexes like flower smells, but women like their flowers simple, while men can enjoy sophistication.

Women in their prime like almonds and lavender, while men in the same age group prefer musk and orange blossoms. Among adults neither sex ranks strawberry essence as high as children do, although mature men like fruity smells better than mature women.

Women like the smell of alcohol better than men do, another puzzler since it is the men who do most of the drinking. The explanation, according to Moncrieff, is that beer, wine, whiskey, and rum contain powerful odorants which hide the smell of the alcohol. Vodka does not have these coverups, but whether or not it would be preferred by women is hard to determine, given the social and economic barriers to drinking it straight.

The odorants which men generally like more than women are mock orange, honeysuckle, wild rose, musk ambrette, ilang-ilang, and lemongrass. Those that women like better than men, in addition to alcohol, naphthalene, and almonds, are alpine violet, bay leaf, and onions.

After the age of forty, women no longer like almond essences much more than men do, and their preference for naphthalene is only marginal. Their liking for musk shows a marked decline, while men still like it very much.

Extraverts and introverts also show different preferences. Not surprisingly, extraverts often like the same fragrances that children do—strawberry essence and red rose—while introverts are more inclined to dislike the things children dislike, such as sophisticated flowery smells.

Another variable is time of day. Body rhythms influence likes and dislikes. A woman may love the smell of onions at dinnertime but detest it at bedtime. A man may like the smell of lavender in the morning but hate it while he is drinking beer in the tavern.

One of the biggest influences on the demand for a particular scent is, of course, advertising. The rush to musk in the 1970s was created by frankly sexy advertisements. The desire of young people in the 1960s for a return to nature was capitalized upon by the "natural scents" promoted in cosmetics and usually made in the factory.

Although most fragrances are still given as gifts at Christmastime, more and more men and women are deciding for themselves how they want to smell.

No odor can be described verbally in English in such a way that it can be immediately recognized or summoned up. There is no foolproof classification or description. Our like or dislike of perfume is strictly personal and subjective.

The type of scent one selects is an indication of temperament. Most perfumers think that people who like violet or frangipani are discerning. Some perfumes will soothe the emotions and some will stimulate them. Some are suitable for use with certain clothes or in a particular climate and others are not.

There are basic categories of scents:

Orientals: These have a heavy, sultry, rich, and sometimes spicy fragrance. They often have overtones of musk and sandalwood. Examples would be Guerlain's Shalimar and Dana's Tabu.

Classic florals: A blending of flower scents such as lily of the valley, jasmine, and light rose. For balance and body, they may contain a medley of basic notes such as amber, musk, vetivert, as well as a touch of the aromatic, but there can be no mistaking the smell of flowers. Examples are Joy by Patou and Lanvin's Arpège.

Fruity blends: Blends which give an air of ripeness and the smell of fruits such as strawberry and peach. Ricci's Bigarade was an example of such a scent.

Modern blends: This is a loose term describing perfumes containing aliphatic aldehydes, which when first introduced into perfumery were considered novel, daring, and modern. Examples are Revlon's Intimate and Norell's Norell.

Woodsy (forest) blend: Generally linked with the aroma of freshly cut, dry wood, they are mossy-leafy or resinous. Such blends have masculine connotations and are used in male toiletries. They often have pungent notes of geranium, lavender, fern, or herbs to add earthiness. The most popular in this field is sandalwood.

Green blend: This gives the scent of fresh cut flowers or vines. Violet leaf or methyl heptine carbonate are examples.

Herbal: Very popular in the 1970s, these fragrances have a medicinal or phenolic note combined with grassiness. Essential oils used most often are thyme, hyssop, calamus, chamomile, and other herbs.

Leather: A sweet pungent smokiness popular in the 1700s and still selling well today. The modern scent results from a blending of methyl ionone and oil of birch tar with other synthetic leather smells.

Generally, the Orientals are used at night, the fruity and green blends during the day, and the leather by men, although there are variations in preferences.

Perfumed products for personal use should be purchased when the buyer feels well. Mood can affect the way fragrances smell on the skin. Scent should be selected slowly. A haphazard sniff is not enough.

The first sniff or immediate effect of a fragrance upon the sense of smell is called the "top note." It consists of the volatile part of the perfume, the first impression. It is one of the most important factors in the success of a perfume. The second or main characteristic of the perfume composition is called the "body note." It has a much longer life on the skin and it usually contributes the final stage or dry-out—the foundation of the perfume.

When buying a scent, it is best to shop alone. Friends sometimes influence the decision wrongly by revealing their own personal preferences, not what smells good on the person buying the scent.

It is difficult to evaluate more than three test fragrances at one time. One should not arrive at the counter wearing a perfume or an after-shave lotion, since this will affect the perception of the new scent.

Fragrance should be applied directly to the wrist. A sniff at the bottle is not accurate because the evaporating alcohol masks the true scent. If there is a question about a particular scent, it is best to leave the counter and to return after the new scent has been worn for a while.

When is a scent worth a high price?

Since most fragrances are still bought as gifts, a lot of money is spent on the packaging. Therefore, expensive does not necessarily mean good. A fine, well-blended fragrance should have beauty, overall scent perfection in composition, and diffusion.

The phrase "diffusion of a fragrance" is often used in perfumery. It is frequently confused with strength. The two are not identical. Strength depends more on the type of scent. We say a perfume is weak or strong depending on whether its fragrance is faint or is noticed without effort. Diffusion, on the other hand, is the radiance of the fragrance spreading out by itself into the space surrounding us. A perfume can be very pleasant if you sniff it directly on the skin, but it also has to have the quality of being noticed at a distance. It should even linger for a certain time in the air after you have left the room.

In a good perfume all the ingredients will evaporate at the same rate, so that it will continue to smell the same as long as you wear it. Besides technical quality, a fine fragrance has to have a specific character so it will stay in people's memories.

A fragrance smells different on each user. A perfume is not "finished" until it mixes with the oils of the skin. How it smells depends upon the person wearing it. A perfumer found that his jasmine perfume smelled like a cheap fly spray on a woman who suffered from a thyroid condition.

How long a perfume lasts depends not only on the chemistry of the skin but also on atmospheric conditions. Fragrances, particularly perfumes, should be purchased in small quantities, kept away from light and heat, and tightly capped after each use. As soon as a bottle of perfume is opened, it begins to lose some of its scent; and perfumes, like wine, "suffer" if transported by sea or air.

A particular scent is usually available in many forms—perfume, cologne, after-shave lotion, bath powder, bath oil, and sachet. The concept that cologne is merely diluted perfume and toilet water diluted cologne is a common error. One perfumer explained it by saying that perfume is like a symphony orchestra playing the full masterpiece. Eau de cologne does not contain the rich notes, and toilet water is merely the background of the music.

Better perfumes usually contain between 20 and

24 ounces of oil per gallon of alcohol, but some contain as much as 36 ounces of oil and other products called perfume contain no more than 10 ounces. There are no legal limits.

The more dilute fragrance solution, generally containing 8 ounces of oil per gallon, is called "toilet water" or "eau de cologne" in America, where these designations have been used interchangeably. Historically, the term eau de cologne was used to specify a type of fragrance containing a refreshing citrus note. Nevertheless, any fragrance today which is more dilute than perfume can be called eau de cologne in America, although some of the old European perfume houses still maintain the traditional distinction.

Since there are no legal restrictions, the words "toilet water" and "cologne" have come to mean more dilute and less expensive fragrances. The two are usually sold in relatively large containers at prices far below those of true perfumes. Toilet waters and colognes contain between 3 and 6 ounces of perfume oil per gallon of alcohol with the limits sometimes as low as 2 ounces and sometimes as high as 8.

Another major difference between perfumes and the more dilute forms of fragrance is the concentration or "proof" of the alcohol. A perfume with 16 to 24 ounces of oil per gallon may contain a very low percentage of water in order to remain clear and to have the oil go readily into solution. For this purpose, 95 percent alcohol is usually employed. When the percentage of oil is lower, a weaker alcohol may also be used, usually 80 to 85 percent and sometimes as low as 75 percent. The amount of water present in the alcohol will be determined by the type of oil being used and its solubility characteristics, the percentage of perfume oil per gallon of alcohol, and the demands of the formulation as to the costs. Also considered is the effect obtained with varying percentages of water as far as the fragrance is concerned.

Some perfumers maintain that in addition to a lower cost, watered alcohol gives the fragrance more of a lift, more of a feeling of freshness than concentrated alcohols.

American women have taken to using a wardrobe of perfumes instead of just one. However, George Balanchine, the choreographer, believes that a woman should have a personal perfume. He told an interviewer for *Vogue* magazine that he knows French perfumes by heart and assigns each dancer at the New York City Ballet a perfume according to her identity as he perceives it. In Russia, he said, all the ballerinas have their individual perfumes, and when they are on stage the air smells like a flower garden. He maintains that when he enters an empty elevator he should be able to tell who has just left it. He insists that instead of putting a dab or two of perfume behind each ear the ballerinas must spray it completely over their hair and whole body so they can be smelled as they walk down the hallway.

Perhaps women who are not ballerinas should not go to such extremes, and, indeed, most could not afford to use perfume so profusely. According to industry statistics, women use perfume less than three times a week, on the average, and usually only when going out.

The most effective way of wearing perfume is to spray it on the body after a bath, before dressing. Scent should always be worn below the waist as well as above because fragrance rises.

Perfume is ineffective on clothing and may leave spots. For those who cannot afford perfume, one of the best substitutes is bath oil.

A good perfume should last from three to four hours, although the wearer will not be able to smell it for that long because of the fatigue phenomenon.

If a woman wants to increase her scent sensuality, she should put perfume between her breasts, in the crook of her elbows, and at the back of her knees. A man who wishes to increase his sensuality may rub a scent over the hair on his chest, behind his knees, and in his groin. The warmth of the body diffuses the fragrance, and its discovery by a lover in unexpected places can be very exciting.

Scent and sex are inseparable. Although most researchers working the field are reluctant to identify any one secretion as definitely a human pheromone, the fact is that a whiff of the vaginal scent of fertile women causes sexual arousal in bulls, male goats, and male monkeys. On the other hand, there are legends of perfumed women being pursued by antlered bucks and male beavers. Of course, the perfume industry has long used the products from animals such as musk and civet to create a sexy atmosphere for humans. When commercial perfumes containing these ingredients are put before male dogs, the dogs become sexually aroused.

What about the use of personal perfume for men?

American men are still self-conscious about the use of fragrances even though such use has been heavily promoted since the early 1930s.

Many men still associate cologne with male effeminacy, but there were lightly scented talcs for men as early as the 1920s. Perfumes for men were first promoted in 1937 as "virile," or "refreshing." They had such fragrances as "seasoned leather," "fine tobacco," "old liquor," and "deep cedar forest."

By the late 1930s, Aqua Velva began promoting

perfume for men with a sexy connotation. The single line of copy read: "She noticed the difference in your skin."

Mennen's Skin Bracer was more blatant by the 1940s. The advertisements came right out and said that it was "100 percent male scent."

Men have always liked the scents of pine, musk, jasmine, sandalwood, and lavender. To think they didn't was unrealistic. Women also like male favorites. In fact, today a large percentage of Frenchwomen douse themselves with male colognes because they find the scents stimulating to them as well as to the men.

Fragrant grooming aids for men really began to take off in the mid-1960s and hit the market with such male blockbusters as Hai Karate and Aztec. In fact, the ad for Aztec read "before, and during." Centaur was said by its advertising agency to be the first cologne developed for a man that transmits its virile message only in "moments of close and intimate contact."

Advertisements for men grew even bolder in 1966 when Max Factor told men that they should "arm themselves with Royal Regiment" before "every encounter." Lust, power, potency is the message sold through the scent.

Juanita Byrne-Quinn, the British fragrance market researcher, says: "It is a mistake to think that men don't like perfume. Men use deodorants not as deodorants but as perfumes. Quite specifically, before going to bed with a man, women use perfume as a mood stimulant."

Keeping up with the times, today's advertisements show that men can be tender and far less aggressive than in the past. In pictures showing men holding infants the copy reads, "It's easy to be tough . . . tender takes a lot of doing . . ." The product promoted is a "no-nonsense fragrance for men who are self-assured."

Staid scientists are now proving in the laboratory what the ancients knew instinctively and what Madison Avenue has been heralding for years—perfumes influence the senses of both men and women. Fragrances are still today's chief stimulants for lust and desire aside from human pheromones.

The truth is that we remove our human pheromones by bathing frequently—more than is necessary for health—and then replace them with the pheromones of animals. And we further mask our own odors with other scented chemicals such as mouthwashes, dentifrices, shampoos, deodorants, and soaps.

We influence our own behavior and the behavior of others with both natural and synthetic smells. More and more research is taking place trying to understand this link between odors and behavior. It is obvious that the reactions to a specific smell are spontaneous and immediate. The sudden appearance of an odor can cause measurable changes in the resistance of the skin of a person quite similar to that which takes place if he or she is suddenly startled.

When an odor is liked, there is a relaxation of the facial muscles, smiling, a pleasant tone of voice, laughing, nodding, opening of the mouth, and deeper respiration. When an odor is disliked, there is a turning away of the head and sometimes the entire body. The head may be jerked back, the nose wrinkled, and the upper lip raised. The individual speaks with disgust and makes characteristic sounds such as "ugh" or "phew." There may be coughing, compression of the lips, rubbing of the nose, frowning, putting the hand over the mouth, actual spitting, and a waving away of the source of the smell.

Association with experiences may play a part in determining such reactions, since odors are so closely linked to emotions and memory. Coloring may also affect the sense of pleasantness and unpleasantness. Green coffee would not smell as good as brown, and black strawberries would not seem as fragrant as red.

When it comes to food, successful cooks know that aroma is the difference between an adequate and a great meal. The term "aroma" usually describes a sensation which is somewhere between taste and smell, although taste itself is primarily smell.

In the animal kingdom, the selection of food is rather simple. Eating is a function of the fulfillment of a purely physiological need as well as of the availability of food. But with humans, the decision is rather complex. We eat not only to live, but for a variety of reasons. A shared meal, for instance, can afford excellent social contact—hence the business lunch and the dinner party. Various factors enter into the choice of a meal, including cost and cultural preference, but aroma, sight, and touch are the most basic.

As food becomes scarcer and more of it is manufactured and stored over a long period of time, rather than bought fresh and cooked at home, there will be an increasing need to preserve its natural aromas or add artificial ones. A food must smell fresh. Increased understanding of the connection between human food choices and food odors will be extremely important.

Just as we have to consider preserving and adding aromas to our foods, we have to pay attention to the need for a pleasantly scented environment. As the population increases and industrial and residential sections merge, odor pollution will be a growing problem. Furthermore, the natural pleasant scents in our environment—from trees, flowers, and grasses—will give way to the odors of concrete and chemicals. New products made out of plastics and other synthetics do not smell as nice as those made of wood, cotton, and other naturals.

Under these circumstances, we would do well to imitate the Greeks, who brought the scent of nature indoors. The Greeks had living rooms which opened onto beautiful gardens, where the most fragrant plants were placed near the windows in the belief that the scent had a salutary effect on the occupants of the house. In medieval monasteries, also, the monks planted sweet-smelling herbs near infirmaries for the benefit of their patients.

Those lucky enough to have gardens should consider the scenting of their homes when planting them. Night-blooming flowers can be placed by bedroom windows and day-blooming plants near the kitchen and living-room windows.

Those who live in apartments can bring all sorts of fragrant plants into their homes, which not only beautify but humidify, deodorize, and scent the air with their leaves and blooms.

Lavender has been used for centuries on sheets to encourage peaceful sleep, and a room scented with roses has been used as a "tranquilizer."

Medical attention to odors has just begun. It is believed that some day in the not too distant future, fragrances will be used in medicinal ways for sinus and other respiratory problems as well as for the treatment of the mentally ill. The beneficial effect of a light, airy scent on depressed or fatigued patients has been noted by both physicians and laymen.

The sense of smell is closely allied with the imagination. In 1836 Theophile Thore said in his book *Arts des Parfums* that scents can be as expressive as colors— that while painting and sculpture represent the object directly, perfumes, like music, reveal the intuition of things.

We humans no longer have to hunt for our food or flee through the jungles to save our lives. We can both select a mate and survive without our sense of smell. Nevertheless, our sensitivity to smells remains. There is an undeniable connection between our noses and our drives and emotions.

It may be that in our overeagerness to deodorize we have silenced much of our natural odor communication, but we still converse, knowingly and unknowingly, by scents.

The many scientists who are studying human olfaction today may produce a richer life for us all tomorrow as they develop more understanding about how we communicate with each other by scents and how smells affect our health and behavior. We may begin to really appreciate the most primal of our senses and be conscious of and respond to its myriad messages. We will then be able to use scents to make our lives and the lives of those with whom we associate more enjoyable.

Bibliography

Amoore, John E. "Evidence for the Chemical Olfactory Code in Man." *Annals of the New York Academy of Sciences,* vol. 237 (September 27, 1974), pp. 137–43.

────── and L. Janet Forrester. "Specific Anosmia to Trimethylamine: The Fishy Primary Odor." *Journal of Chemical Ecology* (in press).

──────, ──────, and Ron G. Buttery. "Specific Anosmia to 1-Pyrroline: The Spermous Primary Odor." *Journal of Chemical Ecology,* vol. 1 (July 1975), pp. 299–310.

──────, ──────, and Paolo Pelosi. "Specific Anosmia to Isobutyraldehyde: The Malty Primary Odor." *Chemical Senses and Flavor* (in press).

────── and James R. Popplewell. "Sensitivity of Women to Musk Odor: No Menstrual Variation." *Journal of Chemical Ecology,* vol. 1 (July 1975), pp. 291–97.

Baker, John. *Race.* New York: Oxford University Press, 1974.

Bedrick, Roy. *The Sense of Smell.* Garden City, N.Y.: Doubleday & Company, 1960.

Bennett, Marvin H. "A Reversible Nasal Block for the Rat." *Physiology and Behavior,* vol. 7 (1971), pp. 269–70.

──────, School of Medicine, University of Pennsylvania. Personal communication with author, April 4, 1975.

Berglund, Birgitta. "Quantitative and Qualitative Analysis of Industrial Odors with Human Observers." *Annals of the New York Academy of Sciences,* vol. 237 (September 27, 1974).

Berglund, Ulf. "Dynamic Properties of the Olfactory System." *Annals of the New York Academy of Sciences,* 1972, pp. 17–27.

Bienfary, Ralph. *The Subtle Sense.* Norman, Okla.: The University of Oklahoma Press, 1942.

Breisacher, Peter. "Neuropsychological Effects of Air Pollution." *American Behavioral Scientist,* 1974, pp. 837–64.

Bruce, Hilda M. "Olfaction." *Journal of Reproductive Fertility,* Suppl. 19, 1973, pp. 403–4.

———. "Pheromones." *British Medical Bulletin,* vol. 26, no. 1 (1970), pp. 10–13.

Byrne-Quinn, Juanita. "Perfume—A Tool or a Toy?" *Drug and Cosmetic Industry,* September 1974.

Caroom, Doug, and F. H. Bronson. "Responsiveness of Female Mice to Preputial Attractant Effects of Sexual Experience and Ovarian Hormones." *Physiology and Behavior,* vol. 7 (1971), pp. 659–62.

Cheal, Mary Lou, and Richard L. Sprott. "Social Olfaction: A Review of the Role of Olfaction in a Variety of Animal Behaviors." *Psychological Reports,* 1971, pp. 215–43.

Cohen, I. Kelman, M.D., Paul J. Schechter, M.D., Ph.D., and Robert I. Henkin, M.D., Ph.D. "Hypogeusia, Anorexia, and Altered Zinc Metabolism Following Thermal Burn." *The Journal of the American Medical Association,* vol. 223 (February 19, 1973), pp. 914–16.

Comfort, Alex. "The Likelihood of Human Pheromones." In *Pheromones,* ed. by Martin Birch. Amsterdam: North-Holland/American Elsevier, 1944.

Cone, Thomas E., Jr., M.D. "Diagnosis and Treatment: Some Diseases, Syndromes, and Conditions Associated with an Unusual Odor." *Pediatrics,* vol. 41, no. 5 (May 1975), pp. 993–95.

Coniglio, L., and L. G. Clemens. "Stimulus and Experimental Factors Controlling Mounting Behavior in the Female Rat." *Physiology and Behavior,* vol. 9 (1972), pp. 263–67.

Connolly, F. H., and N. L. Gittleson. "The Relationship Between Delusions of Sexual Change and Olfactory and Gustatory Hallucinations in Schizophrenia." *British Journal of Psychology,* vol. 119 (1971), pp. 443–44.

Crawshaw, Ralph, M.D. "Oh Where Is the Balm of Gilead?" *Prism,* January 1976.

Bibliography 161

———. "The Bedpan Factor." *Prism*, October 1975.
Davies, Jane V., and D. Bellany. "Effects of Female Urine on Social Investigation in Male Mice." *Animal Behavior*, vol. 22 (January 1974), pp. 239–41.
Davis, Richard G. "Acquisition of Verbal Associations to Olfactory Stimuli of Varying Familiarity and to Abstract Visual Stimuli." *Journal of Experimental Psychology: Human Learning and Memory*, vol. 104, no. 2 (March 1975).
———. "Olfactory Psychophysical Parameters in Man, Rat, Dog, and Pigeon." *Journal of Comparative and Physiological Psychology*, vol. 85 (1973), pp. 221–32.
"Deodorants & Antiperspirants." Pamphlet prepared for the AMA Committee on Cutaneous Health and Cosmetics, 1970.
Devor, Marshall, and Michael Murphy. "The Effect of Peripheral Olfactory Blockade on the Social Behavior of the Male Golden Hamster." *Behavioral Biology*, vol. 9 (1973), pp. 31–42.
Distelheim, Irving H., M.D., and Andrew Dravnieks, Ph.D. "A Method for Separating Characteristics of Odors in Detection of Disease Processes 1." *International Journal of Dermatology*, vol. 12, no. 4 (July–August 1973).
Doty, Richard L. "A Cry for the Liberation of the Female Rodent: Courtship and Copulation in Rodentia." *Psychological Bulletin*, vol. 81 (March 1974).
———. "An Examination of Relationships Between the Pleasantness, Intensity, and Concentration of 10 Odorous Stimuli." *Perception & Psychophysics*, vol. 17, no. 5 (1975), pp. 492–96.
———. "Odor Preferences of Female Peromyscus Maniculatus Bairdi for Male Mouse Odors of P. M. Bairdi and P. Leucopus Noveboracensis, a Function of Estrus State." *Journal of Comparative and Physiological Psychology*, vol. 81, no. 2 (1972), pp. 191–97.
———. Personal interview with author, Philadelphia, Pa., June 11, 1975.
Dravnieks, Andrew, Ph.D. "Evaluation of Human Body Odors: Methods and Interpretations." Paper presented at the

December 1974 meeting of the Society of Cosmetic Chemists, New York.

———, L. Keith, B. K. Krotoszynski, and J. Shah. "Vaginal Odors: GLC Assay Method for Evaluating Odor Changes." *Journal of Pharmaceutical Sciences*, vol. 63, no. 1 (January 1974).

Engen, Trygg. "The Effect of Expectation on Judgments of Odor." *Acta Psychologica: European Journal of Psychology*, vol. 36 (1972), pp. 450–58.

———. "The Potential Usefulness of Sensations of Odor and Taste in Keeping Children Away from Harmful Substances." *Annals of the New York Academy of Sciences*, vol. 237 (September 27, 1974), pp. 224–28.

——— and Bruce M. Ross. "Long-Term Memory of Odors With and Without Verbal Descriptions." *Journal of Experimental Psychology*, vol. 100, no. 2 (October 1973), pp. 221–26.

Erb, Russell C. *The Common Scents of Smell.* Cleveland: World Publishing Co., 1968.

Fleming, Alison, and Jay Rosenblatt. "Olfactory Regulation of Maternal Behavior in Rats." *Journal of Comparative and Physiological Psychology*, vol. 86, no. 2 (1971), pp. 221–32.

Fortuna, Michael, and Robert Gandelman. "Elimination of Pain-Induced Aggression in Male Mice Following Olfactory Bulb Removal." *Psychology and Behavior*, vol. 9 (1972), pp. 397–400.

"450 Frenchwomen Can't Be Wrong: A Study of the Perfume Preferences, Buying and Usage Patterns of 450 Wealthy Parisiennes." Pamphlet prepared by PPL, 1974.

Freeman, Stanley K., Ph.D. *Odor in Humans.* Union Beach, N.J.: Research Center, International Flavors and Fragrances, Inc., 1972.

Gault, F. P., and D. R. Coustan. "Nasal Air Flow and Rhinencephalic Activity." *Electroencephalography and Clinical Neurophysiology*, vol. 18 (1965), pp. 617–24.

——— and R. N. Leaton, M.A. "Electrical Activity of the

Bibliography 163

Olfactory System." *Electroencephalography and Clinical Neurophysiology*, vol. 15 (1963), pp. 299–304.

Genders, Roy. *Perfume Through the Ages*. New York: G. P. Putnam's Sons, 1972.

Getchell, Thomas V. and Marilyn L. "Signal-Detecting Mechanisms in the Olfactory Epithelium: Molecular Discrimination." *Annals of the New York Academy of Sciences*, vol. 237 (September 27, 1974), pp. 62–75.

Girgis, M. "The Amygdala and the Sense of Smell." *Acta Anatomy*, vol. 72 (1969), pp. 502–19.

Griffiths, Nerys M., and R. L. S. Patterson. "Human Olfactory Responses to 5a-androst-16-en-3-one—Principal Component of Boar Taint." *Journal of the Science of Food & Agriculture*, vol. 21 (January 1970), pp. 4–6.

Henion, Karl E. "Odor Pleasantness and Intensity: A Single Dimension?" *Journal of Experimental Psychology*, vol. 90, no. 2 (1971), pp. 275–79.

Henkin, Robert I., M.D. "Medical Importance of Taste and Smell." *The Journal of the American Medical Association*, vol. 218, no. 11 (December 13, 1971).

———. Personal interview by Eleanor Nealon, Georgetown University, Washington, D.C., January 1, 1976.

Hoye, Robert C., M.D., Alfred S. Ketcham, M.D., and Robert I. Henkin, M.D. "Hyposmia After Paranasal Sinus Exenteration or Laryngectomy." *The American Journal of Surgery*, vol. 120 (October 1970), pp. 485–91.

Hussey, Hugh, M.D. "Taste and Smell Deviations: Importance of Zinc." *The Journal of the American Medical Association*, vol. 228, no. 13 (June 24, 1974).

Jones, R. B., and N. W. Nowell. "The Coagulating Glands as a Source of Aversive and Aggression Inhibiting Pheromone in the Male Albino Mouse." *Physiology and Behavior*, vol. 11 (1973), pp. 455–62.

——— and ———. "The Urinary Aversive Pheromone of Mice: Species, Strain, and Grouping Effects." *Animal Behavior*, vol. 22 (1974), pp. 187–91.

Jung, Frederic, M.D. "Beer and Garlic Sausage–Induced Halitosis: De Gustibus Non Est Disputandum." *The Journal of the American Medical Association*, vol. 235, no. 1 (January 5, 1976), p. 88.

Kaye, Bernard M., M.D. "Hazards of Feminine Hygiene Sprays." *Medical Aspects of Human Sexuality*, July 1971.

Keith, Louis. "A Comparison of the Effect of a Suppository and Two Douches on Vaginal Malodorants." *Archiv fuer Gynaekologie*, vol. 215 (1973), pp. 215, 245–62.

———. "Olfactory Study: Human Pheromones." *Archiv fuer Gynaekologie*, vol. 218 (1975), pp. 203–4.

Kennedy, James M., and Kenneth Brown. "Effects of Male Odor During Infancy on the Maturation, Behavior and Reproduction of Female Mice." *Developmental Psychobiology* (1970), pp. 179–89.

Kerekovic, M. "The Relationship Between Objective Disorders of Smell and Olfactory Hallucinations." *Acta Oto-Rhino-Laryngologica Belgica*, vol. 26 (February 5, 1972).

Kern, Stephen. "Olfactory Ontology and Scented Harmonies: On the History of Smell." *Journal of Popular Culture*, 1973, pp. 816–24.

Kimelman, Bathsheva, and Robert Lubow. "The Inhibitory Effect of Preexposed Olfactory Cues on Intermale Aggression in Mice." *Physiology and Behavior*, vol. 12 (1974), pp. 919–22.

Klopping, Hein L. "Olfactory Theories and the Odors of Small Molecules." *Journal of the Agricultural Food Chemistry*, vol. 19, no. 5 (1971), pp. 999–1003.

Koelega, Harry S., and E. P. Koster. "Some Experiments on Sex Differences in Odor Perception." *Annals of the New York Academy of Sciences*, 1974.

Koster, E. P. *Psychological Dimensions of Odour Perception.* Report of the Psychological Laboratory of the University of Utrecht, the Netherlands, 1975.

Kouwenhoven, T. "Olfactory and Gustatory Problems: An Introduction to the Technological, Nutritional and Physiological Aspects of the Organoleptic Assessments of Food Characteristics." *World Review of Nutrition and Dietetics*, vol. 12 (1970), pp. 318–76.

Bibliography 165

Krames, Lester. "Role of Olfaction Stimuli During Copulation in Male and Female Rats." *Journal of Comparative and Physiological Psychology*, vol. 85 (1973), pp. 528-35.

Largey, Gale Peter, and David Rodney Watson. "The Sociology of Odors." *Mental Health Digest*, vol. 4, no. 10 (October 1972), pp. 36-40.

Law, John H., and Fred E. Regnier. "Pheromones." *American Review of Biochemistry*, vol. 40 (1971), pp. 533-48.

Leaton, Robert N. "Exploratory Behavior in Rats with Hippocampal Lesions." *Journal of Comparative and Physiological Psychology*, vol. 59, no. 3 (1965), pp. 325-30.

Maxwell, Norman. "Men's Fragrances: Old Taboos Have Faded Away." *Drug & Cosmetic Industry*, May 1975.

McKenzie, Dan, M.D. *Aromatics and the Soul*. London: William Heinemann, 1923.

Michael, Richard P. "Hormones and Sexual Behavior in the Female." *Hospital Practice*, December 1975, pp. 69-76.

———. Personal interview with author, Atlanta, Georgia, October 21, 1975.

———, R. W. Bonsall, and M. Kutner. "Volatile Fatty Acids, 'Copulins,' in Human Vaginal Secretions." *Psychoneuroendocrinology*, vol. 1 (1975), pp. 1-11.

———, ———, and Patricia Warner. "Human Vaginal Secretions: Volatile Fatty Acid Content." *Science*, December 27, 1974.

Moncrieff, R. W. *Odour Preferences*. London: Leonard Hill, 1966.

Motokizawa, F., and N. Furuya. "Neural Pathway Associated with the Egg Arousal Response by Olfactory Stimulation." *Electroencephalography and Clinical Neurophysiology*, vol. 35 (1973), pp. 83-91.

Mulder, D. J. "Dogs in Diagnosis." *The Lancet*, September 4, 1971, p. 555.

Mybytowgcz, R. "Reproduction of Mammals in Relation to Environmental Odours." *Journal of Reproduction and Fertility*, vol. 19 (December 1973), pp. 433-46.

Nadler, R. D., Ph.D. "Sexual Cyclicity in Captive Lowland Gorillas." *Science*, vol. 189 (September 5, 1975).

Nahum, Louis H. "Olfactory Communication: A Human Pheromone?" *Connecticut Medicine*, vol. 35, no. 6 (1971).

"National Institutes of Health: The Politics of Taste and Smell." *Science*, vol. 187 (January 17, 1975).

National Survey of the Odor Problem: Phase I of a Study of the Social and Economic Impact of Odors. Prepared for the National Air Pollution Control Administration by Copley International Corporation, January 1970.

Nesbitt, Paul D., and Girard Steven. "Personal Space and Stimulus Intensity at a Southern California Amusement Park." *Sociometry*, vol. 37, no. 1 (1974), pp. 105–15.

Neuhaus, Walter, and David M. Goldenberg. *Vegetative Reactions After Olfactory Stimulation*. Technical report, Erlangen-Nuremberg University, Erlangen, West Germany, July 1969.

"Odor of Sanity." *The Journal of the American Medical Association*, vol. 212, no. 3 (April 20, 1970), pp. 472–73.

Odors and Air Pollution: A Bibliography with Abstracts. National Technical Information Service, U.S. Department of Commerce, October 1972.

Olfactory Perception. Pamphlet, 3M Company, 1975.

Pangborn, Rose Marie. "Human Parotid Secretion in Response to Pleasant and Unpleasant Odorants." *Pathophysiology*, vol. 10, no. 3 (May 1973), pp. 231–36.

Peckham, Brian W. "Some Aspects of Air Pollution: Odors, Visibility, and Art." Presentation made at a seminar, "Economics of Air and Water Pollution," sponsored by the Water Resources Research Center, Virginia Polytechnic Institute, Blacksburg, Virginia, April 28–30, 1969.

Pefiffer, C. C., and V. Iliev. "Pyroluria, Urinary Mauve Factor, Causes Double Deficiency of B-6 and Zinc in Schizophrenics." *Federation Proceedings*, vol. 32, no. 3 (1973).

Poynder, T. Michael. "Response of the Frog Olfactory System to Controlled Odour Stimuli." *Journal Society of Cosmetic Chemists*, vol. 25 (1974), pp. 183–202.

Preti, George, and George Huggins. "Cyclical Changes in Volatile Acidic Metabolites of Human Vaginal Secretions and

Their Relation to Ovulation." *Journal of Chemical Ecology*, vol. 1, no. 3 (1975), pp. 361–76.
Rahaman, Hafeezur. "The Role of the Olfactory Signals in the Mating Behavior of Bonnet Monkeys." *Communications in Behavioral Biology*, vol. 6 (1971), pp. 97–104.
Ralls, Katherine. "Mammalian Scent Marking." *Science*, vol. 171 (February 5, 1971), pp. 443–49.
Rasmussen, Andrew Theodore, Ph.D. *The Principal Nervous Pathways: Neurological Charts and Schemas with Explanatory Notes*. New York: The Macmillan Company, 1947.
Rausch, Rebecca, and E. A. Serafetinides. "Specific Alterations of Olfactory Function in Humans with Temporal Lobe Lesions." *Nature*, vol. 255 (June 12, 1975).
Ray, Charles D., ed. *Medical Engineering*. Chicago: Year Book Medical Publishers, 1974.
Redgrove, H. Stanley. *A Scent and All About It*. Easton, Pa.: Chemical Publishing Co., 1928.
Rivlin, Richard S., M.D. *Impairment of Taste and Smell in Hypothyroid Patients*. Prepared for Eighth Annual Vitamin Information Bureau Seminar, Chicago, October 9, 1975.
Rosebury, Theodor. *Life on Man*. New York: Viking Press, 1969.
Roth, James L. A., and Michael D. Levitt. "The Problem of Gas in the Gut." *Medical World News*, April 21, 1975.
Rovee, Carolyn Kent. "Olfactory Cross-Adaptation and Facilitation in Human Neonates." *Journal of Experimental Child Psychology*, vol. 13 (1972), pp. 368–81.
Russell, Michael J., University of California Medical Center, San Francisco. Personal communication with author, May 1976.
Sakellaris, Peter C. "Olfactory Thresholds in Normal and Adrenalectomized Rats." *Physiology and Behavior*, vol. 9 (1972), pp. 495–500.
Schapiro, Shawn, and Manuel Salas. "Behavioral Response of

Infant Rats to Maternal Odor." *Physiology and Behavior*, vol. 5 (1970), pp. 815–17.

Schechter, Paul J., and Robert I. Henkin. "Abnormalities of Taste and Smell After Head Trauma." *Journal of Neurology, Neurosurgery, and Psychiatry*, vol. 37, no. 7 (July 1974), pp. 802–10.

Schleppnik, Alfred A. Personal communications with author, December 1975, March 1976.

Schneider, Robert A. "Newer Insights into the Role and Modifications of Olfaction in Man Through Clinical Studies." *Annals of the New York Academy of Sciences*, vol. 237 (September 27, 1974), pp. 217–23.

Seegal, Richard F., and Victor H. Denenberg. "Maternal Experience Prevents Pup-Killing in Mice Induced by Peripheral Anosmia." *Physiology and Behavior*, vol. 13 (1974), pp. 339–41.

Shiftan, Ernest. "Civilization and Perfumery." Presented at CIDESCO Conference, August 25, 1972.

———. "New Phase in Perfumery." Presented before Morishita Party (Japan), New York, August 23, 1974.

———. Personal interview with author, New York, 1975.

———. "Philosophy of Perfumery." Presented at CIDESCO 3rd National Congress of Esthetics, New York, March 17, 1975.

———. "The Facts Behind Modern Perfumery." For William D. Witter, Inc., April 24, 1974.

Sinclair, John G. "Reflections on the Role of Receptor Systems for Taste and Smell." *International Review of Neurobiology*, vol. 14 (1971), pp. 159–71.

Sinclair, Robert. *Essential Oils: The Basis of Nature's Perfumes*. London: Unilever Limited, Unilever House, 1957.

Slotnick, Burton. "Olfactory Stimulus Control Evaluated in a Small Animal." *Olfactometer, Perceptual and Motor Skills*, vol. 39 (1974), pp. 583–97.

———. Personal communication with author, June 24, 1975.

——— and Howard Katz. "Olfactory Learning–Set Formation in Rats." *Science*, vol. 185 (August 30, 1974), pp. 796–98.

Smith, Kathleen, M.D. Personal communication with author, April 21, 1975.
——— and Jacob O. Sines, Ph.D. "Demonstration of a Peculiar Odor in the Sweat of Schizophrenic Patients." *A.M.A. Archives of General Psychiatry,* vol. 2 (February 1960), pp. 184–88.
Sorrentiano, Sandy, Jr., Russel Reiter, Don Schalch, and Robert J. Donfrio. "Role of the Pineal Gland in Growth Restraint of Adult Male Rats by Light and Smell Deprivation." *Neuroendocrinology,* August 1971, pp. 116–24.
Speer, Frederic, M.D. "Dietary Allergy in Vascular Headache." *The Journal of the American Medical Association,* vol. 232, no. 4 (April 28, 1975), p. 400.
Sprott, Richard L. "'Fear Communication' via Odor in Inbred Mice." *Psychological Reports,* vol. 25 (1969), pp. 263–68.
Steiner, J. E. "Discussion Paper: Innate, Discriminative Human Facial Expressions to Taste and Smell Stimulation." *Annals of the New York Academy of Sciences,* vol. 237 (September 24, 1974), pp. 229–33.
Stoller, Leonard. "Evolution in Fragrance." *Aerosol Age,* vol. 9, no. 8 (August 1974).
Strauss, Elsa Lovitt, M.A. "A Study on Olfactory Acuity." *Annals of Otology, Rhinology, and Laryncology,* vol. 79 (February 1970).
Svare, Bruce, and Ronald Gandelman. "The Stimulus Control of Aggressive Behavior in Androgenized Female Mice." *Behavioral Biology,* vol. 10 (1974), pp. 447–59.
Tanabe, Teruhisa, Masae Iino, Yuriko Ooshima, and Sadayuki F. Takagi. "An Olfactory Area in the Prefrontal Lobe." *Brain Research,* vol. 80 (1974), pp. 127–30.
Thiessen, Delbert D. "Footholds for Survival: A Study of the Mechanisms Controlling Scent Marking in the Mongolian Gerbil Is Helping Us to Understand the Adaptive Value of Territoriality." *American Scientist,* vol. 61 (May–June 1973).
Truex, Raymond C., Ph.D., and Malcolm B. Carpenter, M.D.

Strong and Elwyn's *Human Neuroanatomy*. Baltimore: Williams and Wilkins Co., 1964.
A User's Guide to Perfumery. Booklet published by Proprietary Perfumes Limited, Ashford, Kent, England, 1974.
Verrill, A. Hyatt. *Perfumes and Spices*. Boston: L. C. Page Co., 1940.
Von Koschembahr, John C. "New Styles in Corporate Management." *Finance*, March 1974.
Wallace, Patricia, Keith Owen, and D. D. Thiessen. "The Control and Function of Maternal Scent Marking in the Mongolian Gerbil." *Physiology and Behavior*, vol. 10 (1973), pp. 463–66.
Wasserman, Edward A., and Donald D. Jensen. "Olfactory Stimuli and the 'Pseudo-Extinction' Effect." *Science*, vol. 166 (December 5, 1969).
Weaver, Elmer R. *Control of Odors*. National Bureau of Standards Circular 491, U.S. Department of Commerce, April 17, 1950.
Whitten, W. K. "Genetic Variation of Olfactory Function in Reproduction." *Journal of Reproductive Fertility*, Suppl. 19, 1973, pp. 405–10.
Winans, Sarah S., and J. Bradley Powers. "Neonatal and Two-Stage Olfactory Bulbectomy: Effects on Male Hamster Sexual Behavior." *Behavioral Biology*, vol. 10 (1974), pp. 461–71.
"A World of Flavors and Scents." *Monsanto Magazine*, First Quarter, 1973.

Index

Acromegaly, 61
Addison's disease, 82–83
Adrenal hormones, 82–83
Africans, odor of, 41
Air fresheners, 138
Air pollution, 125–131
Air swallowing, 63, 66
Alcohol: and breath odor, 62–63; odor of, 145; in perfumes, 150–151
Alcoholism and cacosmia, 81
Almond essence, 144, 145
Ambergris, 92, 93, 96, 117
American Museum of Natural History, New York, 123
American plants brought to Europe, 95
Americans, perfumes used by, 111, 150, 151, 152
Amino acids, 64
Ammonia, 64, 137
Amoore, John, 31–32, 42–43
Amygdala, 29, 30
Anesthetics as deodorizing agents, 136–137
Animals: mate selection, 53–54; as mothers, 54; perfumes affecting, 152; pets, odors of, 131; scent glands, 41–42, 53, 67; and scent of fear, 68–69; sexual odor, 34, 42–44, 47–54; throat and mouth in sense of smell, 79; *see also* Estrus cycle
Anosmia, 77
Antony, 91
Apocrine glands, 37, 38, 40–42, 52; of animals, 41–42, 53, 67

Arabs, 27, 35; perfumes used by, 92, 95
Arthritis, 82
Asafetida, 143
Ashurbanipal, 89–90
Assyria, perfumes in, 89–90
Atlanta, Ga., air pollution in, 125
Attar of roses, 93
Automobiles, scent in interior of, 107
Avicenna, 93

Baker, John, 41
Balanchine, George, 151
Bathing, 16, 93, 154
"Bedpan factor," 58
Bible, perfumes mentioned in, 88–89, 91–92
Birds, mating, 53
Birth-control pills, 50–51, 66
Bishop, Hazel, 120
Blacks: nose structure of, 25; odor of, 40, 41
Body odor, 13–17; and age, 42; and disease, 57, 61–62, 66–68; emotions affecting, 68–69; identification by, 21–22, 35, 41; male and female, 38–39; in mental disorders, 69–70; physiology of, 36–38; racial or cultural, 35–36, 40–41; and sex, *see* Sexual odors
Borneol, 143
Braille Institute of America, 123
Brain: electrical response to odors, 29–31, 47; identification of odors, 30; in olfactory system, 24, 27–31

171

Brain tumors, effect on sense of smell, 76, 78
Breath odor, 62–63; in disease, 59–60; and food, 62, 134
Bromidrosis, 59
Butyl mercaptan, 105
Butyric acid, 127
Byrne-Quinn, Juanita, 108–111, 153

Cacosmia, 78, 81
Cagliostro, Count, 101–102
Calcium chloride, 132
California Newspaper Publishers' Association, 123
Camphor, 143
Cancer: odor of, 59, 134; surgery for, and loss of sense of smell, 78–80
Candida albicans, 66
Carbon, activated, 133–134
Carbon dioxide in intestinal gas, 64
Carteret, N.J., chemical plant, 128
Castor, 118
Caswell-Massey, 102
Catherine de Medici, 96–97
Charles II of England, 100
Charles IX of France, 97
Chesebrough-Pond, 107
Children: discrimination of odors, 139; preferences in scents, 144; scents in education of, 123–124
Chimpanzees, sexuality, 49–50
China, incense in, 90
Chlorine, 137
Chlorophyll, odor of, 145
Cilia, 26
Civet, 93, 95, 100, 118, 152
Clayton, Thomas, 101
Clement VII, Pope, 96
Cleopatra, 91
Coal smoke, 127
Cologne, 98, 149–150; for men, 152–153
Color, odor related to, 155
Cone, Thomas E., Jr., 60
Copulins, 50–52
Cosmetics, 117; history of, 91, 100; regulation of materials in, 120
Coty, François, 103
Crawshaw, Ralph, 58, 84
Creosote, 132, 136
Crocodiles, sense of smell, 79
Cromwell, Oliver, 100
Crusaders, 93–94

Darwin, Charles, 25, 33–34, 53

Deodorants, 36, 116, 135–139; in feminine hygiene products, 122–123; in home, 134; malodor counteractants, 137–139; odorless, 136; in underwear and socks, 122
Dimethyl sulfide, 126
Diseases: and loss of taste or smell, 73–77, 81–83; odors of, 57–72, 134–135; scents as protection from, 101
Disraeli, Benjamin, 61
Dogs, 42, 52, 68, 91, 152; sense of smell, 28, 67, 72; use in detection, 72
Doty, Richard, 51
Dravnieks, Andrew, 71–72
Du Barry, Mme, 98, 102
Dysosmia, 78

Earwax, 40–41
Eau de cologne, 98, 149–150
Eccrine glands, 37, 40
Education, scents used in, 123–124
Egypt, ancient, perfumes in, 88
Eleanor, Queen, 127
Electric eels, 72
Elizabeth, Queen of Hungary, 94–95
Elizabeth I, 99–100, 127
Elkton, Md., air pollution in, 128
Ellis, Havelock, 68
Emotions: odor changes in, 68–69; sweat in, 37, 68
Engen, Trygg, 139
England: perfumes, history of, 99–102; perfumes, use of, 111
Enzymes, 31
Epilepsy, 29
Eskimos, 23, 35–37
Estrogen: experiments with animals, 48, 49; treatment with, 45, 46, 75
Estrus cycle: of chimpanzees, 49–50; of gorillas, 50; of mice, 43–44; of monkeys, 48–50
Ethyl mercaptan, 126, 128
Europeans (Caucasians): odor of, 36–37, 40–41; preferences in scents, 142–143

Family odor, 54, 55
Faregeon, 99
Farina, Jean Antoine, 98
Fatty acids, 50–51, 64
Fear, scent of, 68–69
Feet, odor of, 67–68

Index 173

Feminine hygiene deodorants, dangers of, 122–123
Feminis, Jean Paul, 98
Fireflies, 53
Fish, 69; mating, 53
Flatus, 15, 63–66
Fleischman's Distilling Corporation, 107
Fliess, Wilhelm, 34, 35
Flower scents: in perfumes, 147; preferences for, 142, 144–146
Flowers, indoors, 135, 156
Food: artificial scents for, 122; and body odor, 36; and breath odor, 62, 134; and intestinal gas, 64–65; and loss of taste or smell, 78, 81; odor of, 26–27, 155; odor of, imitated, 122, 123; sniffing (testing), 20–21
Food and Drug Administration, 119, 120, 122
Formaldehyde, 137
France: perfumes, history of, 96–99, 103–104; perfumes, use of, 111, 153
Francis I, 96
Francis II, 97
Frankincense, 92
French Boarding House Syndrome, 44–45
Freud, Sigmund, 15–16, 59
Fruit flies, 53
Fruit scents: in perfume, 147; preference for, 144, 145

Galopin, Auguste, 35
Gama, Vasco da, 95
Genital areas, 37, 39–41
George III, 102
Georgetown, D.C., air pollution in, 129
Germany, 109, 110; perfumes used in, 111
Girls: delayed menstruation, 75; isolated from men, menstrual cycles, 44–45
Glyoxal, 136
Gonadal system, 75–76
Gorillas, sexuality, 50
Greece, ancient: houses in, 156; perfumes in, 91
Guerlain, Jacques, 103–104

Haeckel, Ernst, 34
Hair: odors in, 134; in underarm and genital areas, 38, 40

Hallucinations of smell, 70, 78
Halston, 104
Head injury and loss of sense of smell, 77
Health, Education and Welfare Department, 130
Hemophilus vaginalis, 66
Henkin, Robert I., 46, 74–83
Henry II of France, 96
Henry III of France, 94, 97
Henry IV of France, 97
Henry VIII of England, 99
Hepatitis, 74
Herbal (natural) scents, 146, 147
Heterosmia, 78
Hildegard of Bingen, 94
Hippocrates, 63, 91
Home odors, 131–135; control of, 132–135, 138
Horses, 68, 91
Hospitals, smell of, 135
Household products, scent of, 106, 109–110
Huebner, Darrel, 124
Hungary water, 94–95, 98
Hydrogen in intestinal gas, 64–65
Hydrogen sulfide, 64, 126, 127, 128
Hyperhidrosis, 59
Hyperosmia, 82–83
Hyposmia, 78–80
Hypothalamus, 48

Incense, 90, 92, 135
India, 95; perfumes used in, 90, 92
Indole, 64
Industrial plants: odor control in, 130; odors from, 125–131
Infants: breast and bottle feeding, 16; intestinal gas, 65; odors in disease, 60–61; sense of smell, 29, 55
Influenza, loss of taste or smell after, 77
Insects, mating, 53
International Flavors and Fragrances, 103, 112, 114, 120, 121, 123
Intestinal disorders, breath in, 60, 62
Intestinal gas: composition of, 63–65; passing of, 15, 63–66
Ionizers, 137
Isaac and Jacob, 35
Isobutyraldehyde, 32
Isovaleric acid, 32
Isovaleric acidemia syndrome, 60–61

Italy, perfumes made in, 96, 100

Jaeger, Gustav, 34, 35
Japan, incense in, 90
Japanese, 143; odor of, 36, 40, 41; perfumes used by, 142
Jasmine, 102, 153
John the Good, 94
Josephine, Empress, 99
Jovan, 104
Judith and Holofernes, 89

Kallman's Syndrome, 47–48
Keller, Helen, 21
Kelly, Smelly, 20
Kennedy, John F., 82–83
Kidd, Capt. William, 102
Kissing, 35, 40
Krafft-Ebing, Richard von, 16
Krotoszynski, Boguslav, 71–72

Laird, Donald, 105
Laryngectomy, 80
Lauder, Estée, 104, 111
Laundry products, scented, 106, 109–110, 119
Lavender, 144, 145, 153, 156
Lavender water, 94
Leather, odor of, 96, 107, 147
Le Magnen, J., 45–46
Levitt, Michael, D., 63–66
Lilly, Charles, 101
Limbic system, 28, 29
Lin, A. Oscar, 121
Linnaeus, Carolus, 42
Loss of sense of smell, 73–84; classification of disorders, 77–78; diseases in, 73–77, 81–83; surgery related to, 78–80
Louis XIII, 97
Louis XIV, 94, 97–98
Louis XV, 98, 102
Louis XVI, 98
Lysol, 132, 136

Magellan, Fernando, 95
Malodor counteractants, 137–139
Malodors, see Odors, unpleasant
Maori, 35
Maple-syrup urine disease, 61
Marguerite de Valois, 97
Marie Antoinette, 98
Mary Stuart, Queen of Scots, 97
Maugham, W. Somerset, 33
Memory of odors, 22–23, 136, 142
Menopause, 46, 77

Menstrual blood, 42–43; secretions in, 50–51
Menstruation: of girls isolated from men, 44–45; nose in, 34; odor in, 43; and sense of smell, 45–46; and zinc deficiency, 75
Mental disorders: body odor in, 69–70, 72; hallucinations of smell in, 70, 78; sense of smell in, 69
Mercaptans, 126
Methane in intestinal gas, 64, 65
Mice, 54; sexuality, 43–44
Michael, Richard, 50–51
Michael's Mixture, 51–52
Middle Ages, perfumes used in, 93–94
Middle East, perfumes used in, 90, 92, 93
Mohammed, 92
Mohammedans, 23, 92
Monardes, Nicolas, 95
Moncrieff, R. W., 143–145
Mondeville, Henri de, 94
Monilia, 66
Monkeys, sexuality, 48–50
Morris, Desmond, 55–56
Moses, 89
Mosques, 92
Mouth, sense of smell in, 79
Muhammad, 92
Musk: in mosques, 92; odor of, 45, 46, 144, 145; in perfumes, 52, 88, 93, 96, 99, 100, 117–118, 143, 146, 152, 153; synthetic, 118
Musk deer, 117–118
Myrrh, 88, 89, 92

Naphthalene, 132, 144, 145
Naphthenic acid, 126
Napoleon, 99
Natural scents (herbal), 146, 147
Natural selection, odor in, 53–54
Nero, 92
New York City, air pollution in, 128
New York University, Research Center for Mental Health, 124
Nitrogen in intestinal gas, 63, 64
Nose: anatomy, 23–26; cultural symbolism, 23; functions, 26–29; odor of discharge from, 61–62; and sexual behavior, 34
Nose brain, 28–29
Nose rubbing, 23, 35

Nylon: stockings, scented, 105–106; underwear, 66–67

Odor sensory cells, 27–28
Odors: adaptation to, 30–31; allergy to, 83–84; and behavior, 154; brain reactions to, *see* Brain; diagnostic studies of, 70–72; of disease, 57–72, 134–135; in environment, 155–156; importance of, 141–157; memory of, 22–23, 136, 142; national and cultural preferences in, 143–144; preferences, age and sex differences in, 144–145; primary, 32; as warning signals, 73–74, 139–140; *see also* Body odor; Breath odor; Perfumes; Scents; Sexual odors
Odors, unpleasant, 125–140; in air pollution, 125–131; in homes, 131–134, 138; masking of, 107, 135–139; *see also* Deodorants
Olfactory bulb, 27, 29, 30, 75; absence of, 47–48
Olfactory system, 23–32
1-pyrroline, 32
Orange blossoms, 144, 145
Orange-flower water, 98, 100
Orientals: odor of, 36–37, 40–41; preferences in scents, 142–143
Oxford, Edward de Vere, Earl of, 99–100
Oxygen in intestinal gas, 63, 64
Ozena, 61–62
Ozone, 137

Paper, scented, 123
Paper mills, air pollution by, 125, 126, 128
Peckham, Brian W., 127–128
Pen, scented, 121
Perfume industry, 102–104; scents in advertising and selling, 105–124, 146
Perfumes, 36, 56, 87–124; categories of, 146–147; choice of, 110–111, 115; diffusion of, 148–149; forms of, 149–150; history of, 87–104; ingredients of, 117–118; men's preferences in, 111, 144–145, 152–153; national preferences in, 111, 116, 142–143; "noses" in, 112, 114; regulation of materials used, 119–120; selection, principles of, 148–149;

sex and temperament in preferences, 144–146; and sex appeal, 152–154; strength of, 120, 148–149; synthetic materials in, 118–119; use, technique of, 151–152
Perkins School for the Blind, 123
Peru, Indians of, 41
Phantosmia, 78
Phenols, 126
Phenylketonuria, 60
Pheromones, 18–19, 32, 39, 42, 46–47, 152, 154; in animals, 43–44, 46, 49, 52
Philippe Auguste, 94
Philippine Islanders, 35
Pine scent, 135, 153
Pituitary gland, 61; hormones, 48, 61
Plague, scents as protection from, 101
Poison gas, 105
Poisons, 96
Polo, Marco, 95
Pompadour, Mme de, 98
Popcorn, odor of, 107
Poppaea, 92
Portugal, 95
Preti, George, 51
Primary odors, 32
Procter and Gamble, 106
Proprietary Perfumes, Ltd., 103, 108, 109, 116
Puberty and sense of smell, 75
Putrescine, 127

Rabbits, sexuality, 52
Rats, 54; scent of fear, 69; sexuality, 47, 48
Red chypre, 93
Renaissance, perfumes used in, 94–97
René the Florentine, 96
Respiration, 25–26
Rhinencephalon, 28
Rivlin, Richard S., 81–82
Rome, ancient, perfumes in, 91–92
Rose water, 99
Roth, James L. A., 63–66
Ruggiero, 96
Russell, Michael J., 38–40, 55

Saddle Brook, N.J., air pollution in, 129
Saint-Exupéry, Antoine de, 103–104
Samoans, 35

Sandalwood, 146, 147, 153
Scent glands, see Apocrine glands
Scents: in advertising and selling, 105–124, 146; categories of, 146–147; on clothing, 102, 105–106, 121–122; in education, 123–124; in environment, 155–156; "fresh," 107, 109; healing powers of, 84, 101, 156; "natural," 146; on newspapers, 123; as protection from disease and evil, 101; scratch-and-sniff, 106–107, 123–124; on underwear, 102, see also Odors; Perfumes
Schizophrenia, odors in, 69–70, 72, 78
Schleppnik, Alfred A., 137–138
Scratch-and-sniff, 106–107, 123–124
Sebaceous glands, 37, 38
Sebum, 37, 38
Sex hormones, 46–47; deficiencies and olfactory disorders, 80; in Kallman's Syndrome, 48
Sexual odors, 16, 33–56; of animals, 34, 42–44, 47–54; male and female, 38–39
Sexuality: and loss of sense of smell, 73, 75–77; perfume associated with, 152–154
Shakespeare, William, 100
Sheba, Queen of, 89
Shiftan, Ernest, 112–116, 119
Sickroom, odors in, 134–135
Sinuses, surgery on, 78–79
Sjögren's disease, 82
Skatole, 64, 127
Skunk smell, 105, 126
Smell, see Odors; Perfumes; Scents
Smell, sense of: importance of, 17–19, 141–157; in language, 14–15; loss of, 73–84
Smell This Shirt, 121–122
Smith, Kathleen, 69–70
Snakes, odor of, 69
Sniffing (investigation): of food, 20–21; of people, 71, 72
Soap, 99, 100; scented, 106, 119, 120
Socks, deodorized, 122
Soda, baking, 134
Solomon, 89
Solon, 91
Spain, perfumes made in, 95–96, 104
Spanish leather perfume, 96

Spices, trade in, 93, 95
Steggerda, F. R., 64
Stockings, scented, 105–106
Strawberries, scent of, 107, 144, 145, 146
Sweat: in bromidrosis, 59; and emotions, 68; of feet, 68; in hyperhidrosis, 59; medications for control of, 68; and color, 37–38
Sweat glands, 37, 46

Taste, sense of, 27, 155; loss of, 73–78, 81–82
Testosterone, 44, 45, 46, 75
Theophrastus, 127
Thomas, Lewis, 45
Thore, Theophile, 156
3M Company, 106, 123, 124, 139
Thyroid hormone deficiency, 81–82
Ticonderoga, N.Y., paper mill, 128
Tobacco smoke, 131
Toilet water, 149–150
Trichomonas vaginalis, 66
Trigeminal nerve, 30
Trimethylamine, 32, 42–43
Turbinates, 25–26
Turpentine, 137
Tuscany, Duke of, 102

Underarm areas, 37–41
Underwear: deodorized, 122; nylon, 66–67; scented, 102, 122
Urine: of animals, 43–44, 49, 52–53; incontinence, control of odor, 134–135; odor in disease, 60–61

Vaginal glands, 46
Vaginal infections, 66–67
Vaginal secretions, 50–51; odor of, 66–67, 71–72, 152
Valerian, 142–143
Vanilla, scent of, 144–145

Walter, Henry, Jr., 120–121
Washing products, see Laundry products, scented; Soap
Washington, George, 102
Watkins, Hugh, 116–117, 119
Weber, Albert, 20–21, 83
Wegener's granulomatosis, 82
Wells, H. G., 33
Whitten, W. K., 44
Wicomico County, Md., air pollution in, 129–130
William III, 127

Zinc and sense of smell, 74–76